Communications
in Computer and Infor   82

Achim Gottscheber
David Obdržálek   Colin Schmidt (Eds.)

Research and Education in Robotics - EUROBOT 2009

International Conference
La Ferté-Bernard, France, May 21-23, 2009
Revised Selected Papers

 Springer

Volume Editors

Achim Gottscheber
SRH University of Applied Sciences
Electrical Engineering, Heidelberg, Germany
E-mail: achim.gottscheber@fh-heidelberg.de

David Obdržálek
Charles University, Faculty of Mathematics and Physics
Malostranské náměstí 25, 11800 Praha, Czech Republic
E-mail: david.obdrzalek@mff.cuni.cz

Colin Schmidt
ParisTech Angers-Laval
Laboratoire Presence et Innovation, Arts et Métiers, Ingénierium
4 rue de l'Ermitage, 53000 Laval, France
E-mail: colin.schmidt@univ-lemans.fr

Library of Congress Control Number: 2010936712

CR Subject Classification (1998): I.2, I.2.9, I.4, I.5, J.4, I.6

ISSN   1865-0929
ISBN-10   3-642-16369-6 Springer Berlin Heidelberg New York
ISBN-13   978-3-642-16369-2 Springer Berlin Heidelberg New York

springer.com

© Springer-Verlag Berlin Heidelberg 2010
Printed in Germany

Typesetting: Camera-ready by author, data conversion by Scientific Publishing Services, Chennai, India
Printed on acid-free paper   SPIN: 06/3180   5 4 3 2 1 0

Preface

This volume contains the accepted papers presented during the International Conference on Research and Education in Robotics – EUROBOT Conference 2009, held in La Ferté-Bernard, France, May 21–23, 2009.

Today, robots are indispensable tools for flexible, automated manufacturing in many areas of industry as well as for the execution of sophisticated or dangerous tasks in the nuclear industry, in medicine and in space technology, and last but not least, they are being increasingly used in everyday life.

To further encourage research in this area, the EUROBOT Conferences have been set up. They aim to gather researchers and developers from academic fields and industries worldwide to explore the state of the art. This conference is accompanied by the EUROBOT Contest Finals, an international amateur robotics contest open to teams of young people. During the finals in 2009, teams from 25 countries came together not only to compete, but also to exchange knowledge and ideas and to learn from each other. In addition to the paper and poster presentations, there were two invited talks:

- Raja Chatila, Director of the LAAS – CNRS, Toulouse, France
  whose talk was about "Cognitive Robots"
- Véronigue Raoul, EUROBOT Association, France
  whose talk was about "EUROBOT"

Organizing a conference is a task that requires the collaboration of many people. We personally would like to warmly thank all members of the EUROBOT Conference 2009 program committee; without their help and dedication it would not have been possible to produce these proceedings. Their effort deserves special thanks. And of course, many thanks go to all authors who submitted their papers, no matter whether their papers were accepted or not.

The conference was supported by the town of La Ferté-Bernard, France, the Charles University in Prague, Czech Republic, and the Arts et Métiers ParisTech Laboratory (LAMPA), France. Submitted papers and the proceedings have been prepared using the EasyChair system, for which we would like to thank its developers.

May 2009

Achim Gottscheber
David Obdržálek
Colin T. Schmidt

Organization

EUROBOT 2009 was organized by the EUROBOT Association, the Charles University in Prague and the Arts et Métiers ParisTech Laboratory (LAMPA), Paris, France, in cooperation with the town of La Ferté-Bernard, France and Planète Sciences Association, France.

Executive Committee

Conference Chair	Achim Gottscheber (SRH University of Applied Sciences, Heidelberg, Germany)
Program Chair	David Obdržálek (Charles University in Prague, Czech Republic)
Local Organization	Colin Schmidt (Le Mans University and LAMPA, Arts et Métiers ParisTech, France)

Program Committee

Jacques Bally	Y-Parc, Parc Scientifique et Technologique, Yverdon-les-Bains, Switzerland
Kay Erik Böhnke	Heidelberg University of Applied Sciences, Heidelberg, Germany
Branislav Borovac	University of Novi Sad, Serbia
Jean-Daniel Dessimoz	Western Switzerland University of Applied Sciences, HESSO HEIG-VD, Yverdon-les-Bains, Switzerland
Heinz Domeisen	Hochschule für Technik Rapperswil, HSR/IMA, Rapperswil, Switzerland
Boualem Kazed	University of Blida, Algeria
Pascal Leroux	ENSIM/LIUM, Le Mans University, France
Julio Pastor Mendoza	Universidad de Alcal, Madrid, Spain
Giovanni Muscato	University of Catania, Catania, Italy
Fernando Ribeiro	Universidade do Minho, Guimaraẽs, Portugal
Simon Richir	LAMPA, Arts et Métiers ParisTech, France

Table of Contents

Walk-Startup of a Two-Legged Walking Mechanism

Kalman Babković, László Nagy, Damir Krklješ, and Branislav Borovac

Faculty of Technical Sciences
Trg Dositeja Obradovića 6, 21000 Novi Sad, Serbia
bkalman@uns.ac.rs

Abstract. There is a growing interest towards humanoid robots. One
of their most important characteristic is the two-legged motion – walk.
Starting and stopping of humanoid robots introduce substantial delays.
In this paper, the goal is to explore the possibility of using a short unbalanced state of the biped robot to quickly gain speed and achieve the
steady state velocity during a period shorter than half of the single support phase. The proposed method is verified by simulation. Maintainig
a steady state, balanced gait is not considered in this paper.

1 Introduction

Today, there is a growing interest towards humanoid robots. These are similar
to humans in many aspects - measures, shape, locomotion, vision, etc. Among
all characterisics, one of the most important is the two-legged motion – walk.
Because of this one characterictic, such robots can easily adapt to human environments. Stairs, uneven surfaces, grooves etc. would represent an unavoidable
obstacle to a wheeled robot.

In the recent years, there are numerous implementations of two legged robots.
Some of them use the well known ZMP theory to achieve dynamic balance during
their walk [1]. Although they can reach a considerable velocity while in steady
state walk, starting and stopping still introduce substantial delays and imbalance
[2]. It should also be noted that the foot, its shape and maneuverability plays
a significant role in biped walking. Flat feet are the classical approach to biped
walk implementation, but there are situations when it is impossible to maintain
full contact of the flat foot sole with the ground [3].

In this paper, the goal is to explore the possibility of using a short unbalanced
state of the biped robot to quickly gain speed and achieve the steady state
velocity during the course of one quarter of a step (half of a single support
phase). To accomplish this, a very fast action is itroduced at the ankle of the
walking mechanism (figure 1). In other words, the actuator acts on the shank
forcing it forwards in a very short interval, but because of the inertia of the
system, the front of the foot rises and the heel is pressed against the ground.
In turn the whole mechanism rises and the leg remains almost parallel to itself.
After that, since the mechanism is not balanced anymore (only the foot edge at

A. Gottscheber, D. Obdržálek, and C. Schmidt (Eds.): EUROBOT 2009, CCIS 82, pp. 1–10, 2010.
© Springer-Verlag Berlin Heidelberg 2010

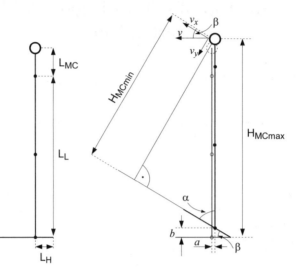

Fig. 1. Lateral view of the walking mechanism's legs – the dots represent the joints, the hollow circle is the position of the main mass center. **left:** initial position, **right:** after the sudden foot lift.

the heel is in contact with the ground), the mechanism commences to overturn forward, gaining a substantial forward velocity and can possibly enter the steady state walk. This situation will be referred to as walk-startup.

It has to be noted, that the mechanism should soon come into balance again by extending the other leg forward. If this doesn't happen, the mechanism will eventually fall. In this paper, that kind of action will not be considered. The possibilities of the abovementioned walk-startup will be examined both analitycally and by simulation.

2   Results Obtained Analytically

The simplified schematics of the walking mechanism is given in figure 1. The total mass of the mechanism is $65kg$ and it is located at the top of the mechanism. In the following analysis it is considered that the mass of the legs and feet can be neglected. The exact dimensions (marked in figure 1) are given in table 1. β can be also smaller and greater than the β_0 given in table 1. β_0 is the angle used in this analysis. Also, it has to be pointed out that that the mass at the top of the mechanism represents the mass of a potential robot body (including the head, hands, control units, power supply etc).

An estimation of the expected forward-velocity of the mechanism after the startup-phase can be calculated. Looking at the figure 1, assuming that the leg remains parallel to itself, the following can be concluded:

$$\alpha = \frac{\pi}{2} - \beta \tag{1}$$

Table 1. Mechanical parameters of the mechanism

L_H	$8cm$
L_L	$1m$
L_{MC}	$20cm$
mass (total)	$65kg$
β_0	$15°$

The mass reaches its maximal vertical position (2) right after the sudden movement in the ankle is introduced. Right after that movement stops the overturning commences. At the end of the overturing phase, the mass is at its minimum height (3). The overturning phase is considered finished when the foot is parallel with the ground surface.

$$H_{MCmax} = L_{MC} + L_L + L_H \cos \alpha \tag{2}$$

$$H_{MCmin} = (L_L + L_{MC}) \sin \alpha \tag{3}$$

Considering that the initial potential energy ($E_p = mg\Delta H = mg(H_{MCmax} - H_{MCmin})$) is fully converted to kinetical energy ($E_k = mv^2/2$) at the end of the overturnig interval the following can be concluded:

$$v = \sqrt{2g(H_{MCmax} - H_{MCmin})} \tag{4}$$

The horizontal and vertical components of this velocity can be written as

$$v_x = v \cos \beta \tag{5}$$

$$v_y = v \sin \beta \tag{6}$$

The horizontal component (5) is of particular interest because it is close to the overall walking (or forward) velocity of the walking mechanism. It should be noted that the velocities do not depend on the mass at all. In the simulations later in this paper, the situation when $\beta = \beta_0 = 15°$ is analyzed. For that angle and the mechanical parameters from table 1 the above formula (4) gives the result $1.09m/s$ as the center of mass velocity, while the horizontal component (5) is $1.05m/s$.

3   Structure of the Mechanism Used in the Simulation

The situation described earlier was modeled by a computer based numerical model. The mechanism was described as a three-dimensional mechanical structure given in figure 2. The thin vectors inside the mechanism segments represent the vectors pointing from the joints to the mass centers of each segment. All joints are rotational ones. Although the mechanism incorporates many joints, actually only two of them were used in the simulations. These two joints are specially pointed out in figure 2. The other joints are kept in a fixed, immobile postition. The reason why that many joints are introduced is that this way

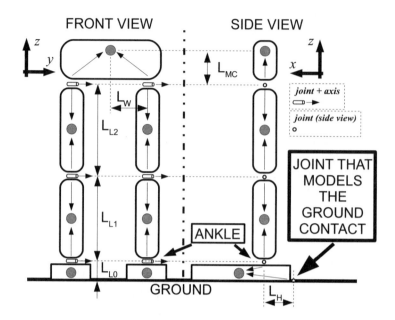

Fig. 2. The mechanism as modeled on a computer

the configuration of the mechanism can be quickly changed without modifying the mechanism definition. This is expected to be useful in further simulation experimentation with the model. All the relevant dimensions were chosen to correspond to the schematic given in figure 1. All elements shown in figure 2 have a non-zero mass, although the mass of the legs is by far smaller than the mass of the segment at the top. All mechanical parameters are given in table 2.

It hast to be noted that since this model is three-dimensional, during the interval when the mechanism is supported by only one foot, there can be a tendency of the mechanism to overturn to its side. This situation is not considered in this simulation. It can be assumed for example, that the support foot is always wide enough to prevent the overturning to the side.

4   Simulation Results

In simulations, the process of foot actuation can be precisely controlled and its effect on the system can be thorougly examined. In the following simulations the angle between the shank and the foot is changed in a predefined manner as if there was a regulator that controls the angle with great accuracy. Different movements are defined – the net change of the angle is always 15°, but the interval during which this change takes place varies. The angular velocity is, howevwer, not constant. There is always an acceleration phase at the beginnig and a deceleration phase at the end. The duration of both phases is expressed as a percentage of the whole angle change interval.

Table 2. Mechanical parameters of the mechanism – simulation analysis

L_{L0}	$8cm$
L_{L1}	$0.5m$
L_{L2}	$0.42m$
L_{MC}	$20cm$
L_W	$15cm$
L_H	$8cm$
M_{upper}	$65kg$
M_{thigh}	$1kg$
M_{shank}	$1kg$
M_{foot}	$0.5kg$

The simulations were conducted using a mathematical model of mechanical systems. It can be used for modeling multibody mechanical systems where all the bodies are connected by rotational joints. The movement in the ankle is predefined and fed directly to the model. There is no additional control system and accordingly no actuator properties were taken into account.

In figure 3 the changing of this angle in time is shown with dashed lines. 4 different cases are shown with different change durations. The solid lines represent the angle between the ground and the foot. It can be easyly noticed that at the beginnig these angles almost coincide (the front end of the foot lifts up from the ground), but as the overturning commences, this angle falls back to zero. The simulation ends when the zero angle is reached.

The next figure (fig. 4) shows the horizontal position of the mass center and its horizontal velocity component in time. The results are given for the same movement of the ankle as depicted in figure 3. One can observe that when the angle change at the ankle is rapid, the overturning of the mechanism is postponed, while for slower changes the mass center advancings almost coincide. If the angle changes too slowly the overturning can be faster than the change in the ankle and therefore the maximum velocity is never reached (thick line in figure 4).

In figures 5 and 6 the vertical and horizontal components of the force exerted to the ground are presented. A very important note must be made here: the ground contact of the mechanism is implemented via a rotational joint in this simulation alnalysis, which means that the mechanism actually cannot loose its contact with the ground. This can only be accepted if the force exerted to the ground is directed downwards – in terms of simulation results, the vertical component must be less than zero. Otherwise, the mechanism would make a jump in reality. In other words, the simulation results are valid only if the resulting vertical force is negative, which is fulfilled in the presented cases. A similar thing holds for the horizontal component – in that case the force may not exceed the statical friction force (that actually depends on the friction coefficient and the vertical component of the force).

It is clear that the forces exerted to the ground are greater if the angle change at the ankle is faster. The force is the greatest during the acceleration phases, and

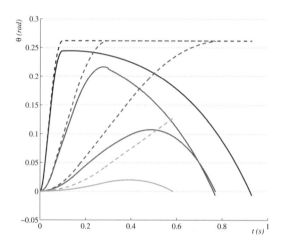

Fig. 3. The Angle between the foot and the shank for different movement durations (**dotted line**) and the foot and the ground (**solid line**) for different angle change duration times – from left to right: $100ms$,$300ms$,$800ms$ and $1.4s$. The acceleration percentage is 20% in each case.

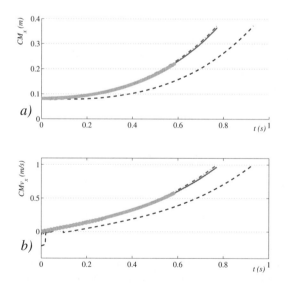

Fig. 4. The horizontal position of the mass center of the system (a) and the horizontal component of its velocity in time (b). Different lines correspond to different angle change duration times: dashed line-$100ms$, dash-dot line-$300ms$,solid line-$800ms$ and thick line-$1.4s$.

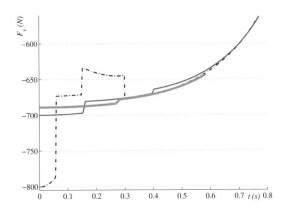

Fig. 5. Vertical force exerted to the ground by the mechanism. Different lines correspond to different angle change duration times: dash-dot line – $300ms$, solid line – $800ms$ and thick line – $1.4s$.

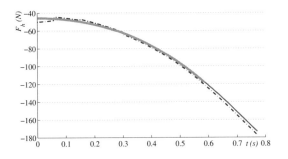

Fig. 6. The horizontal force component exerted to the ground by the mechanism. Different lines correspond to different angle change duration times: dash-dot line – $300ms$, solid line – $800ms$ and thick line – $1.4s$.

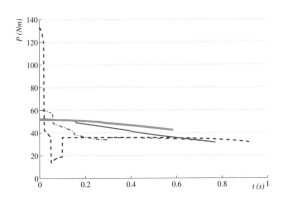

Fig. 7. Actuator torque needed for the angle change at the ankle. Different lines correspond to different angle change duration times: dashed line – $100ms$, dash-dot line – $300ms$, solid line – $800ms$ and thick line – $1.4s$.

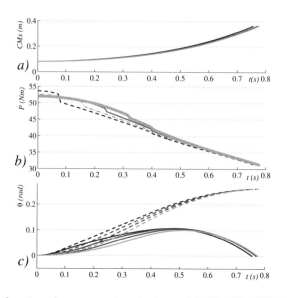

Fig. 8. The acceleration phase percentage is changed (10%,20%,30%,40%) while the net change duration remains $800ms$ – (**a**)Center of mass movement (horizontal position), (**b**) Actuator torques, (**c**) Angle between the foot and the shank (dashed) and the angle between the foot and the ground changing in time.

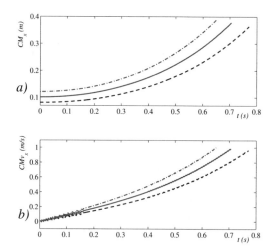

Fig. 9. Center of mass movement (horizontal position (**a**) and horizontal velocity (**b**)) when the net change duration remains $800ms$, the acceleration percentage is 20%, and the heel to ankle horizontal distance changes (L_H in figure 2): dashed:$8cm$, solid:$10cm$, dash-dot:$12cm$.

the smallest during the deceleration phases. When no acceleration ,deceleration or movement is present, the force depends on the overturning angle. The more the mechanism overturns, the smaller the vertical component and the greater the horizontal component becomes. This means, that the mechanism would slip at a certain instant when the statical friction limit decreases below the horizontal force intensity.

In figure 7 the torque generated by the actuator needed to achieve the angle change at the ankle at the desired rate is shown. The greatest torques are needed during the acceleration phases, while during the deceleration phases the torque needed even decreases considerably. When high ratio reduction gears are used, actually no motror torque is needed when the andgle does not change because of the self-braking effect of such mechanisms.

It is clear now that if the angle change duration time is too short the torque required from the actuator can be too high - if the actuator is a DC motor, very high currents are needed for high torques and the high current puts a heavy load on the power supply. The rotor windigns can overheat, or the reduction gear can be damaged, etc. On the other hand, the final velocity of the mass center is reached even slower if the acceleration percentage is too low.

It turns out that the best effects can be achieved if the foot is lifted at a rate similar to the overturnig rate. This way the actuators will not be unnecessarily put under heavy load, but the needed mass center velocity will be reached in the shortest possible time. This situation is achieved when the angle between the foot and the ground remains as close to zero as possible (ex. the third curve from the left in figure 3).

Another interesting question is, what would happen if only the acceleration percentage changed, while the change duration and decelration percentage remained the same. Figure 8 shows the results of this similation experiment.

Since the distance between the projection of the mass center to the ground and the ground contact point of the mechanism is the main overturing tendency measure of the system, it is interesting to explore the consequences of changing this distance by changing the distance between the ankle and the heel (L_H in figure 2). The results are presented in figure 9. It is clear from the figure that when this distance is greater, the desired steady state velocity is reached sooner. An explanation for this is that the starting overturning action is greater when this distance is bigger. As the mechaism overturns, this action increases further in each case. On the other hand, the final CM velocity is almost the same in each case.

5   Conclusion

The walk-startup was sucessfuly analyzed both analytically and by computer simulation. Both analyses resulted in almost the same terminal, steady state velocity. It can be concluded that the steady state velocity can be reached in the interval shorter than one single-support phase. Further investigation can be done towards the possibilities of walk continuation after the startup phase and also towards a model that could handle more cases than the one presented in this paper.

Acknowledgemet

This paper is one of the results of the research project: 114-451-00759/2008, financed by the Provincial Secretariat for Science and Technological Development, Autonomous Province of Vojvodina, Republic of Serbia.

References

1. Vukobratovic, M., Borovac, B.: Zero-moment Point – Thirty Five Years of Its Life. International Journal of Humanoid Robotics (2004)
2. Zhu, C., Tomizawa, Y., Luo, X., Kawamura, A.: Biped Walking with Variable ZMP, Frictional Constraint, and Inverted Pendulum Model. In: IEEE International Conference on Robotics and Biomimetics, ROBIO 2004, August 22-26, pp. 425–430 (2004)
3. Hashimoto, K., Sugahara, Y., Ohta, A., Sunazuka, H., Tanaka, C., Kawase, M., Lim, H., Takanishi, A.: Realization of Stable Biped Walking on Public Road with New Biped Foot System Adaptable to Uneven Terrain. In: The First IEEE/RAS-EMBS International Conference on Biomedical Robotics and Biomechatronics, BioRob 2006, February 20-22, pp. 226–231 (2006)

Control Methodologies for Endoscope Navigation in Robotized Laparoscopic Surgery

Enrique Bauzano, Victor Muñoz, Isabel Garcia-Morales, and Belen Estebanez

Departamento de Ingeniería de Sistemas y Automática, University of Malaga,
Parque Tecnologico de Andalucia
Severo Ochoa, 4 29590 Campanillas (Malaga), Spain
ebauzano@uma.es

Abstract. This paper is focused on the motion control problem for a laparoscopic surgery robot assistant. This article studies both, navigation problem and efforts applied with an actuated wrist when the endoscope interacts with the patient: gravity, friction and abdominal effort. In this way, a new control strategy has been proposed based on a previous work with a passive wrist robotic assistant. This strategy is able to emulate the behavior of a passive wrist by using two nestled control loops. Finally, the paper concludes with a comparison between both wrists, as well as future applications for actuated wrists related to minimally invasive surgery.

Keywords: Surgical robot, Motion control, Force Control.

1 Introduction

Minimally invasive surgery aims to reduce the size of the incisions in order to lessen patients' recovery time and limit any post-operative complication [1]. The disadvantage of this kind of surgery lies in the form of new constrains and difficulties for the surgeon. In particular, the fact of accessing to the abdominal cavity with special long instruments through little incisions in the skin. It produces movement limitations, loss of touch and 3D perception, and hand-eye coordination problems. In this situation, the idea of providing surgeons assistance came about, in the form of new procedures based on robotic engineering.

One of the challenges to be faced by the use of these technologies consists in replacing the assistant surgeon used in laparoscopic surgery procedures for moving the camera. In this type of technique, it is essential for the assistant to focus the camera on the area of interest for the main surgeon. The endoscope may come into contact with a tissue, fail to focus on the area required by the surgeon or transmit unstable pictures to the monitor due to the assistant's steady hand.

The solution for the above-related question is to design a robotic assistant for laparoscopic surgery. This robotic assistant performs the *laparoscopic navigation*, which consists in positioning the endoscope inside the abdominal cavity, in order to show the desired anatomical structure. These movements are contained in a sphere, centered at the endoscope point of insertion (*fulcrum point*), which introduces a holonomic motion constraint. This situation is detailed in Fig. 1.

A. Gottscheber, D. Obdržálek, and C. Schmidt (Eds.): EUROBOT 2009, CCIS 82, pp. 11–22, 2010.
© Springer-Verlag Berlin Heidelberg 2010

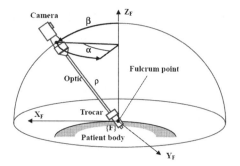

Fig. 1. Endoscopic navigation problem

As it can be noticed from the mentioned figure, the fulcrum point has a task frame attached in order to specify the camera relative location through spherical coordinates α (*orientation angle*), β (*altitude angle*) and ρ (*external distance*, defined as the distance from the wrist to the fulcrum point along the endoscope) [2], [3]. Angles α and β can be calculated by reading the robot's internal sensors, although this does not apply to the value of ρ.

Robot assistant must be equipped with a wrist able to develop these spherical movements as shown in Fig. 1. There are two possible solutions: *passive* or *active* wrists. Passive wrists consist of a mechanism with no actuated joints [4]-[6]. This feature guarantees that no efforts are applied to the abdominal wall of the patient. However, any uncertainty concerning the fulcrum knowledge decreases the endoscope position precision.

On the other hand, the active wrist scheme is based on actuated joints. Therefore, these kinds of wrists are able to locate the endoscope in a more accurate way than passive ones. Nevertheless, since the planned robot motion depends on the fulcrum location, any error in this measure would produce undesired forces applied to the patient's abdomen. In order to solve this issue, some works propose the use of remote rotation centre mechanisms [7], [8]. This paper proposes a different approach based on a direct actuated wrist scheme [9]-[11].

In this way, section 2 reviews the control strategy for passive wrists in laparoscopic navigation. Section 3 describes the proposed methodology by using direct actuated wrists, whereas section 4 focuses on the design of its control strategy based on the previous experience with passive wrists. Section 5 establishes a comparative analysis between the movement of the laparoscopic camera with both passive and active wrists. Finally, section 6 concludes the advantages of using direct actuated wrists versus passive ones, as well as its possible future applications related with minimally invasive surgery.

2 Previous Work: Motion Control with Passive Wrists

The adaptive Cartesian controller system detailed in this section (Fig. 2) has been designed to compensate orientation and insertion errors. This scheme uses a Cartesian trajectory planner, which ensures that the robot's wrist makes the spherical shaped

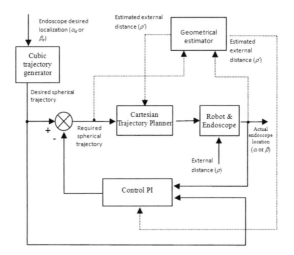

Fig. 2. Adaptive Cartesian controller for passive wrists

movement following a first-order system response, with a constant time τ and a static gain K. In this way, this planner simplifies the complexity of the robot dynamic behaviour, and therefore makes easier the controller design [12].

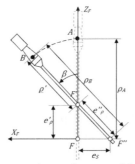

a) Movement computed around a wrong fulcrum point (F')

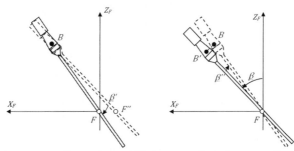

b) Fulcrum correction action   c) Endoscope orientation correction action

Fig. 3. Active wrist navigation problem (a) and the proposed control strategy (b and c)

In this way, a PI control law has been added to the Cartesian trajectory planner. The overall controller, using the desired endoscope location, generates a smooth spherical trajectory that is used as the PI control law reference [13]. The control loop provides feedback of the real location of the endoscope in order to calculate the required spherical trajectory, thus eliminating the location error caused by the uncertainty of the *external distance* ρ'. The Ackerman's methodology for poles assignment has been used to design the PI adaptive control law, according to a dead-beat strategy.

Finally, the mission of the geometric estimator represented in Fig. 2 is twofold: i) adjust the *PI* controller in order to maintain the control specifications, and ii) update the exterior distance estimation ρ, which is used by the trajectory generator.

3   Robot Active Wrist Navigation Problem

This section describes the navigation problem of the laparoscopic camera to the robot-actuated wrist. This configuration avoids the backlash introduced by the trocar-endoscope interaction and the endoscope always reaches the commanded spherical position in spite of an inaccurate estimation of the *external distance* ρ. Nevertheless, the active wrist would force the patient's abdominal wall if the planned movement is not computed around the actual fulcrum point. This situation is detailed in Fig. 3.

Fig. 3 (a) presents an endoscope movement in order to reach the spherical reference coordinate β from a null altitude. This arc shape movement starts from position A and the robot planner computes the final location B by using the estimated fulcrum position F' as the turning centre, instead of the actual one F. Both F and F' are placed along the endoscope axis, however the first one is defined by the *actual external distance* ρ_A and the second one by the *estimated external distance* ρ'. The estimation error is the distance from the related points and it is labelled in Fig. 3 (a) by e'_ρ. Finally, because of the named estimation error e'_ρ, the planned endoscope movement produces a displacement of the actual fulcrum point from F to F'' a distance e_s by forcing the abdominal wall. The collateral effects are: a change in the actual external distance from ρ_A to ρ_B, and the estimation error reaches the value e''_ρ.

The proposed control strategy is designed in other that the endoscope reaches the goal spherical coordinate in the presence of an *external distance estimation error* without forcing the abdominal wall. This strategy is implemented by using the actions described in Fig. 3 (b) and (c). The first action (Fig. 3 (b)) emulates the behaviour of a passive wrist by turning an angle β' around the final position B, in order to eliminate the fulcrum displacement e_s and make coincident F'' with F. During this action, the controller estimates a more accurate value for the external distance. Finally, the desired altitude β of the endoscope is corrected by turning an angle β'' around the actual fulcrum point F, as it is shown in Fig. 3 (c). In this way, the actions described in Fig. 3 (b) and (c) needs an accurate value for the *external distance estimation* ρ'_B at point B, and the angle β''.

Therefore, the structure of this section is divided in two subsections. The first one is devoted to present an endoscope-abdominal wall interaction model useful to estimate the distance e_s. The second subsection details the estimation of the *external distance* from the camera to the fulcrum point. Both subsections assume that the robot has a force sensor installed between the wrist and the endoscope. In this way, two contributions of the measured effort are compensated: the endoscope weight and the friction forces due to the endoscope-trocar interactions. These assumptions about the sensor measurements compensation will be detailed in section 4.

3.1 Endoscope-Abdominal Wall Interaction Model

The model of the endoscope-abdominal wall interaction defines the behaviour of the fulcrum displacement during the endoscope movement, as shown in Fig. 3 (a). This model establishes a relationship between the distance e_s and the force sensor measurements (a six components force-torque vector). Fig. 4 shows the final spherical position presented in Fig. 3 (a), which includes the forces and torque interaction between the endoscope and the abdominal wall. In this way, the movement around the estimated fulcrum position F' produces a force $\mathbf{F_s}$ due to the displacement of the actual fulcrum along the abdominal wall, and a force $\mathbf{F_r}$ caused by the endoscope-trocar friction. Vectors $\mathbf{F_P}$ and $\mathbf{M_P}$ are the force and torque values read by the force sensor with the endoscope weight compensated, so the gravity force has no effect to the sensor readings. Finally, ρ represents the actual external distance and ψ is the angle from the endoscope axis to the abdominal reaction force $\mathbf{F_s}$.

Force $\mathbf{F_s}$ is a function of the abdominal wall elasticity. Some works use complex models based on finite elements modelling in order to determine its deformation under a given force [14], [15]. However, from the point of view of a feedback control, it is possible to use a simple spring model [16] to establish a relationship between the fulcrum displacement e_s and the abdominal reaction force $\mathbf{F_s}$ as it is shown in expression (1).

$$\left|\mathbf{F_s}\right| = \lambda \cdot e_s \tag{1}$$

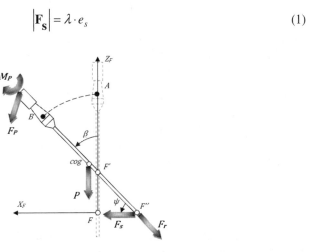

Fig. 4. Endoscope-abdominal wall forces and torques interaction

In the above expression, λ represents the abdominal wall elasticity. By taking into account the null $\mathbf{F_r}$ assumption, the force $\mathbf{F_s}$ is equal to the sensor force reading $\mathbf{F_P}$. On the other hand, λ can be computed by means of the instantaneous variation of the abdominal wall reaction force versus the time:

$$\lambda = \frac{\left|\Delta \mathbf{F_s}\right|}{\Delta t} \tag{2}$$

Expression (2) allows an estimation of the abdominal wall elasticity in each control sample Δt and a means square filter avoids abrupt elasticity changes due to numerical computation imprecision.

Therefore, the combination of expressions (1) and (2) provides an estimation of the value e_s as a function of force $\mathbf{F_s}$ which can be used in a feedback control scheme in order to emulate the behaviour of a passive wrist as it is shown in Fig. 3 (b).

3.2   External Distance Estimator

The external distance ρ is required by the controller in order to execute the β correction movement shown in Fig. 3 (c). Indeed, the passive wrist emulation action (Fig. 3 (b)) changes the desired altitude angle in order to eliminate the abdominal reaction force $\mathbf{F_s}$. Therefore, it is necessary an arc movement around the actual fulcrum position in order to reach again the altitude β. The planning of this movement requires the distance from the fulcrum F to the robot end effector.

An estimation ρ' of this external distance can be obtained by using a momentum balance methodology of the forces and torques shown in Fig. 4. This methodology provides an expression, which connect the abdominal reaction force $\mathbf{F_s}$ with the sensor readings through the external distance ρ. The expression (3) details these relationships where ρ' must be converted to its vector form.

$$\mathbf{M_P} = \rho' \times \mathbf{F_s} + \frac{d}{2} \mathbf{u_s} \times \mathbf{F_r} \tag{3}$$

In the above relationship, $\mathbf{u_s}$ is a unitary vector with the same direction and orientation than $\mathbf{F_s}$, and d is the endoscope diameter. The contribution of the second right side term of this expression is lower than the first one due to the distance ρ' is greater than the endoscope diameter d. Moreover, the compensation for the friction force $\mathbf{F_r}$ is one of the assumptions presented at the beginning of this section. In this way, the second right side element can be eliminated.

By using the above consideration about (3), the estimated distance ρ' can be obtained by applying the module cross product operation to the named expression as it is detailed in the relation (4):

$$\left|\rho'\right| = \frac{\left|\mathbf{M_P}\right|}{\left|\mathbf{F_s}\right| \sin(\psi)} \tag{4}$$

where ψ is the angle from the endoscope axis to the abdominal reaction force vector $\mathbf{F_s}$. Let $\mathbf{u_p}$ be a unit vector that defines the endoscope computed by using the forward robot kinematic, then the expression for ψ is detailed at (5).

$$\sin(\psi) = \frac{\left|u_\rho \times F_s\right|}{\left|F_s\right|} \tag{5}$$

Finally, the external distance estimation ρ' is computed by using the expressions (4) and (5). The feedback controller described in the next section in order to execute the compensation of the spherical coordinates feedback loop will use this information.

4 Robot Active Wrist Control Strategy

Section 3 has proposed a methodology that allows the emulation of a passive wrist with an actuated one. To achieve this goal, the proposed system model makes possible to obtain an estimation of the external distance, as well as the e_s error. This section is focused on the design of a suitable control strategy that solves the endoscope navigation problem by using the parameters ρ' and e_s, described in the previous section. A scheme of this control loop is shown in Fig. 5.

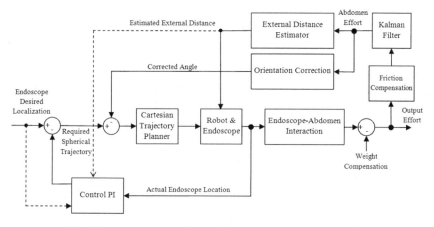

Fig. 5. Endoscope-abdominal wall forces and torques interaction

The proposed scheme consists of two nestled feedback control loops. The inner one is devoted to emulate the behaviour of a passive wrist by feeding back the force and torque readings. The endoscope weight has been compensated, as well as the endoscope-trocar friction and the measurements sensor noise. This loop uses the estimated distance e_s in order to perform the control action described in Fig. 3.b. In this way, the orientation estimator converts the mentioned distance into a compensation angle for the fulcrum correction. On the other hand, the outer loop implements the spherical coordinates control as shown in Fig. 3.c in order to correct the endoscope orientation. This control loop works in the same manner as established for the passive wrist robotic assistant, as shown in section 2 and the Cartesian controller planner has the same function as it was described before.

Firstly, an *offline* test must be done in order to identify the contribution of the endoscope weight. With this information and by using (6), the endoscope gravitation force pairs can be compensated.

$$\mathbf{F_P} = \mathbf{F} - {}^{\{M\}}\mathbf{R}_{\{0\}} \cdot \mathbf{P}$$
$$\mathbf{M_P} = \mathbf{M} - \mathbf{r_{cdg}} \times \left({}^{\{M\}}\mathbf{R}_{\{0\}} \cdot \mathbf{P}\right) \tag{6}$$

The force $\mathbf{F_P}$ and torque $\mathbf{M_P}$ are the computed force and torque with the compensation of the gravity effect, $\mathbf{F}$ and $\mathbf{M}$ are the measured force and torque, all of them related to the reference frame $\{M\}$. On the other hand, $\mathbf{P}$ is the endoscope absolute weight vector, and $^{\{M\}}\mathbf{R}_{\{0\}}$ is the rotation matrix from frame $\{0\}$ to frame $\{M\}$. In case of the torque equation, $\mathbf{P}$ is also multiplied by the center of gravity (*cog*) vector $\mathbf{r_{cog}}$ of the endoscope.

The other non-desirable effort is friction. Its behavior highly depends on the contact surface between endoscope and trocar. During an intervention, this layer may be affected by fluids, therefore friction changes accordingly. Due this unforeseeable variation on this force, it is not an easy task to obtain a model which suits to these changes. However, its direction is known to be over the axis of the endoscope. Moreover, the main interaction between endoscope and trocar is through its internal valve. This element transmits the abdominal effort mainly on a normal direction, so the hypothesis established on this article is that endoscope is inserted perpendicularly into the abdomen. Thereby, measures done by sensor effort over endoscope axis are all due to friction, whereas perpendicular ones are generated by abdomen effort.

To avoid the noise on the sensor measures, it has been decided to add a *Linear Kalman Filter* [17]-[19]. It consists of a simplified model of the real system through a space-state description. Since state of linear, its general expression is shown on (7).

$$x_{k+1} = \left(Ax_k^i + Bu_k\right) + K_k\left(y_k - Cx_k\right) \tag{7}$$

Equation (7) defines the Kalman Filter prediction for the vector state of the real system, expressed on discrete time. Index k means actual instant and $k+1$ the next one. First term on brackets is the linear relation A between actual vector state x and next one, as well as B relates input u. Second term fixes the vector state with an internal feedback of the real system output y with the output modeled by Kalman, also as a linear relation with state x through matrix C. This difference is controlled by a gain K, and it depends on matrices A and C as well as noise estimations.

To define a vector state of the system, Fig. 3 shows that there only exists an independent state parameter: the error on *external distance* e_p. That is, (7) is a one-dimensional expression in this situation, and it turns into (8) for endoscope-abdomen system model, just replacing $A=\cos\beta$ and $B=0$.

$$e_\rho^{k+1} = e_\rho^k \cos\beta^k + K^k\left(F_s^k - \lambda e_\rho^k\right) \tag{8}$$

Real system output is abdominal effort F_s, whereas estimated Kalman output is dependent on abdomen stiffness $C=\lambda$. Once the abdomen effort has been obtained in Fig. 5, the separation error can be calculated, as it was commented on section 3.1. This distance allows to estimate a fix for the initial turning angle, and the feedback of this value makes a movement similar to passive wrists.

The control strategy finishes by means of an external feedback, which turns out to be similar to the loop shown in passive wrists on section 2. The only difference consists of the external distance estimator obtains the information not only from system geometry, but also from the abdomen effort.

5 Implementation and Experiments

In order to verify the proposed controller, it has been necessary to carry out some experiments that probe the appropriate control system operation. A patient simulator has been used for the *in-vitro* procedures, which are required to see how the system responds to artificial, but tangible, tissues.

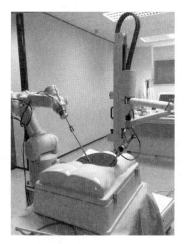

Fig. 6. Robot assistants with actuated wrist (on the left) and passive wrist (on the right) used on experiments

The main purpose for these experiments is to establish a comparative analysis between the movement of the laparoscopic camera with passive wrists and with active ones. In this way, the proposed experiment studies the temporal response on an altitude movement that modifies the β angle from, for example, -48.1 to -43.1 degrees. The controller parameters have been established in order to reach the reference in 3 seconds. Different experiments [20] have shown that a time constant of 0.6 is comfortable for the surgeon. The experiments with passive wrist have been carried out with the ERM robot, which has been fully designed and built in the University of Malaga in order to assist in laparoscopic surgery by moving the camera [21], [22]. On the other hand, an industrial manipulator (PA-10 by Mitsubishi Heavy Industries, Ltd) has been used for the experiments with active wrist (see Fig.6).

5.1 Passive Wrist

The *desired spherical trajectory*, shown in Fig. 7 (a) by a dashed line, is the reference generated via a trapezoidal interpolation for reaching the goal altitude from the current camera position. This trajectory is the entry of the proposed adaptive PI controller (Fig. 5), which feeds back the current endoscope orientation in order to compute the *required β trajectory* (dot-dashed line). Finally, the *actual β trajectory*, obtained via passive wrist encoder readings, is shown as a solid line. As shown, this last trajectory follows the desired one and it reaches the steady state in the goal of β.

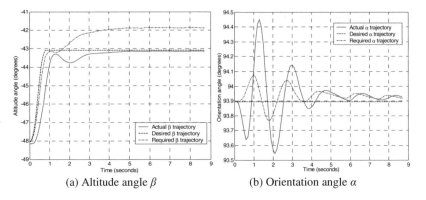

(a) Altitude angle β (b) Orientation angle α

Fig. 7. (a) Temporal response for reaching a desired altitude reference β and (b) Movement error evolution in an altitude movement with active wrist

Fig. 7 (b) represents the orientation angle, α, when the altitude movement is carried out. It can be noticed that, at the beginning of the movement, α coordinate overdamps (with a maximum amplitude of 0.6 degrees). It is due to mechanical imprecision and encoder resolution. Nevertheless, the designed controller compensates for the overdamping and, later, the coordinate keeps its reference value.

5.2 Active Wrist

Fig. 8 (a) details the response of the control system with active wrist proposed in section IV under the same experiment conditions. The represented trajectories have a similar description as the ones with passive wrist in Fig 7 (a). As it can be noticed, the *real β trajectory* follows the *desired* one in an accurate way. Thus, the error evolution for each trajectory has been represented in Fig. 8 (b), in order to show that the differences between them are small: the position error of the altitude angle β never exceeds 0.5 degrees. Moreover, it can be noticed from the mentioned figure that there is not an appreciable delay between the *actual trajectory* and the *desired* one.

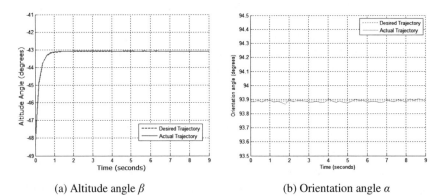

(a) Altitude angle β (b) Orientation angle α

Fig. 8. (a) Temporal response for reaching a desired altitude reference β and (b) Movement error evolution in an altitude movement with active wrist

On the other hand, orientation angle is not involved on movement chosen for the experiment, so its magnitude does not depend on the fulcrum point. Therefore, a comparative analysis of the orientation angle error evolution between Fig. 7 (b) and Fig. 8 (b) with passive and active wrists does not make sense: the robot with active wrist guarantees the tracking of the *desired trajectory* with a null error. That means that the overdamping in the orientation angle with passive wrists is avoided with active ones.

6 Conclusions

This article has described the problem of laparoscopic movement, using robotic assistants with both passive and actuated wrists. More specifically, the control motion on actuated wrists has been developed as an emulation of passive wrist mechanisms. The tool used in both cases has been an endoscope, which just makes efforts on the fulcrum point. After the experimental results, we can conclude that the behaviour on active wrists produces smaller positioning errors than passive ones.

Furthermore, actuated wrists could be more attractive for future applications. Since they are capable of applying efforts, they not only improve the accuracy on location, but also can be used to other more active tasks, for example: holding an organ, stitch operations and, in general, any action required inside the abdominal cavity. However, such tasks apply additional efforts over the tool tip other than the abdominal ones, which must be identified. A passive wrist is unable to apply efforts directly by itself, so it would be less useful and errors committed would be higher.

References

1. Berkelman, P., Cinquin, P., Troccaz, J., et al.: State-of-the-Art in Force and Tactile Sensing for Minimally Invasive Surgery. IEEE Sensors Journal 8(4), 371–381 (2008)
2. Hurteau, R., De Santis, R.S., Begin, E., Gagner, M.: Laparoscopic Surgery Assisted by a Robotic Cameraman: Concept and Experimental Results. In: Proc. of 1994 IEEE International Conference on Robotic & Automation, San Diego, California, USA, pp. 2286–2289 (1994)
3. Pan, B., Fu, Y., Wang, S.: Position Planning for Laparoscopic Robot in Minimally Invasive Surgery. In: Proc. of 2007 IEEE International Conference Mechatronics and Automation, Harbin, China, pp. 1056–1061 (August 2007)
4. Deml, B., Ortmaier, T., Seibold, U.: The Touch and Feel in Minimally Invasive Surgery. In: IEEE International Workshop on Haptic Audio Visual Environments and their Applications, Ottawa, Canada, pp. 33–38 (October 2005)
5. Zemiti, N., Ortmaier, T., Morel, G.: A New Robot for Force Control in Minimally Invasive Surgery. In: Proc. of 2004 IEEE International Conference on Intelligent Robots and Systems, Sendai, Japan (2004)
6. Zemiti, N., Morel, G., Ortmaier, T., Bonnet, N.: Mechatronic Design of a New Robot for Force Control in Minimally Invasive Surgery. IEEE Transactions on Mechatronics 12(2), 143–153 (2007)
7. Berkelman, P., Cinquin, P., Troccaz, J., et al.: A Compact, Compliant Laparoscopic Endoscope Manipulator. In: Proc. of 2002 IEEE International Conference on Robotic & Automation, Washington, DC, pp. 1870–1875 (2002)

8. Lum, M.J.H., Rosen, J., et al.: Optimization of a Spherical Mechanism for a Minimally Invasive Surgical Robot: Theoretical and Experimental. IEEE Transactions on Biomedical Engineering 53(7), 1440–1445 (2006)
9. Zemiti, N., Ortmaier, T., Morel, G.: A New Robot for Force Control in MIS. In: Proc. of 2004 IEEE International Conference on Intelligent Robots and Systems, Sendai, Japan, pp. 3643–3648 (2004)
10. Krupa, A., Morel, G., de Mathelin, M., et al.: Achieving high precision laparoscopic manipulation using force feedback control. In: Proc. of IEEE International Conference on Robotics and Automation, Washington, DC, USA, pp. 1861–1869 (May 2002)
11. Michelin, M., Poignet, P., Dombre, E.: Geometrical control approaches for minimally invasive surgery. In: Workshop on Medical Robotics Navigation and Visualization (2004)
12. Muñoz, V.F., García-Morales, I., et al.: Control movement scheme based on manipulability concept for a surgical robot assistant. In: Proc. of 2006 IEEE International Conference on Robotic & Automation, Orlando, FL, pp. 245–250 (2006)
13. Muñoz, V.F., García, I., Fernández, J., et al.: On Laparoscopic Robot Design and Validation. Integrated Computer-Aided Engineering 10(3), 211–229 (2003)
14. Huang, P., Gu, L., Zhang, J., Yu, X., et al.: Virtual Surgery Deformable Modelling Employing GPU Based Computation. In: 17th International Conference on Artificial Reality and Telexistence, Aalborg University Esbjerg, Denmark, pp. 221–227 (2007)
15. Rahman, A., Khan, U.H.: Effect of Force on Finite Element in Vivo Skin Model. In: Second IEEE International Conference on Emerging Technologies (ICET 2006), Peshawar, N-W.F.P, Pakistan, pp. 727–734 (2006)
16. Nakamura, N., Yanagihara, M., et al.: Muscle Model for Minimally Invasive Surgery. In: IEEE International Conference on Robotics and Biomimetics, pp. 1438–1443 (2006)
17. Lu, X., Wang, W., et al.: Robust Kalman Filtering for Discrete-time Systems with Measurement Delay. In: Proceedings of the 6th World Congress on Intelligent Control and Automation, Dalian, China, pp. 2249–2253 (2006)
18. Luca, M.B., Azou, S., et al.: On Exact Kalman Filtering of Polynomial. IEEE Transactions on Circuits and Systems 53(6), 1329–1340 (2006)
19. Tanabe, N., Furukawa, T., et al.: Kalman Filter for Robust Noise Suppression in White and Colored Noises. In: IEEE International Symposium on Circuits and Systems, ISCAS 2008, pp. 1172–1175 (2008)
20. Muñoz, V.F., Gómez-de-Gabriel, J.M., García-Morales, I., Fernández-Lozano, J., Morales, J.: Pivoting motion control for a laparoscopic assistant robot and human clinical trials. Advanced Robotics 19(6), 695–713 (2005); ISSN 0169-1864, VSP Brill Academic Publishers
21. Muñoz, V.F., García, I., Fernández, J., et al.: Adaptive Cartesian Motion Control Approach for a Surgical Robotic Cameraman. In: Proc. of 2004 IEEE International Conference on Robotic & Automation, New Orleans, LA, pp. 3069–3074 (2004)
22. Muñoz, V.F., García, I., Fernández, J., et al.: Risk analysis for fail-safe motion control implementation in surgical. In: World Automation Congress, pp. 235–240 (2004)

Fast Object Registration and Robotic Bin Picking

Kay Böhnke and Achim Gottscheber

University of Applied Sciences, Heidelberg, Germany
kay.boehnke@gmail.com,
Achim.Gottscheber@fh-heidelberg.de

Abstract. Businesses have invested a lot of money into intelligent machine vision, industrial robotics and automation technology. The proposed solution of this paper deals with industrial applications of robotic bin picking. In this paper, a pose estimation approach is introduce to determine the coarse position and rotation of a known object using commonly available image processing tools applied to 3D laser range data. This position and orientation is refined by a combination of the well-known Iterative Closest Points Algorithm with the hierarchical object representation of Progressive Meshes to find objects in a industrial environment. This approach is integrated in an object localization system for industrial robotic bin picking.

Keywords: Robotic bin picking, Object Localization, 3D Registration.

1 Introduction

Imaging a 2-years-old baby taking his favorite toy car out of a box full of toy cars of different shapes, sizes, and colors; everyone can agree with this: This is an easy task for the baby. At first glance, everyone takes this for granted. Everyone is able to learn this in the first years of his life. Unsupervised recognition and localization of an unknown object in a cluttered environment is still an unsolved problem in the field of robot vision. Today robots get more and more involved in industrial processes, because they are superior to man regarding requirements on strength, speed and endurance. Robotic automation processes became very successful in the last years, and offers a wide range for research. This paper deals with the well-known "bin picking problem". It is also known as the depalletizing problem, which occurs in nearly every industrial sector. Robotics has dealt with this task a very long time, but there are only few solutions suitable for special applications. The most critical and time consuming part in the bin picking process is object localization. Many approaches exist to solve this for a small group of objects. With the advances in 3-D scanning technologies efficient and robust algorithms for geometric models are needed in many fields of research. We introduce a two step approach for object localization. After the data acquisition the pose estimation finds some good matching objects in the bin with standard 2D-image processing algorithms. If the position and orientation

A. Gottscheber, D. Obdržálek, and C. Schmidt (Eds.): EUROBOT 2009, CCIS 82, pp. 23–37, 2010.

of these objects are roughly known, the pose refinement finds the exact match of the object.These candidates are matched with an improved Iterative Closest Points (ICP) Algorithm with 3D range data information. The refinement step in the hierarchical system finds the exact transformation of the model and scene data set using Progressive Meshes inside the iterative calculations of the ICP Algorithms. This reduces its complexity by comparing only the reduced representation of an object model to a scene data set. The obvious advantage is an increasing performance. But another profound effect is the improved robustness against outliers. Many algorithms, methods and object representations exist for surface registration problems. Especially in computer graphics object location and pose estimation is a common task. Therefore, this paper describe an application using modern algorithms adapted from computer graphics. There exists a huge variety of applications in architectural, medical, industrial and scientific 3D-visualization, 3D-modeling, reproduction, reverse engineering and 3D-image processing[1]. The performance of surface registration methods, like the Iterative Closest Point (ICP) algorithm [2], mainly depends on the efficiency of the object representation. The main disadvantage of the ICP algorithm is the computational complexity. In most cases this algorithm can not be used to process huge data sets in real time. Many improvements were made in the past to speed up the registration process [1][3]. The most time consuming part of the ICP algorithm is the fact that for each point in one data set the algorithm needs to find the closest point in the other data set. Our goal is to reduce this search time by decreasing the number of iterations and the number of corresponding points in each iteration step. To ensure the efficiency we will focus on the representation of object and the characteristics of the algorithms. The main idea is that we adjust the level of detail of the data sets in each iteration step of the ICP process. These contributions are evaluated in our experiments in section 5. We introduce the basics of the ICP and the Progressive Meshes in section 3 and the combination of both in section 4. Then, we introduce an object localization system in section 6 for robotic bin picking and conclude in section 7.

2   Related Work

Only a small number of papers include the whole process for bin picking. Brady et al. [4] give an overview of solutions concentrating on object localization from the early 80th to 1988. Kak and Edwards [5] surveyed the state of art of bin picking in 1995. They focused on feature based methods and optimal tree search algorithms. One of the early papers related to the bin picking problem was published by Ikeuchi et al. [6] in 1983. Surface orientation histograms are matched to histograms of simulated orientations of CAD-models (represented as unit sphere projections as Extended Gaussian Image (EGI)). These surface orientations are calculated from multiple views of a convex object. An experimental setup evaluated the results with a process time of one minute for a few torus shaped objects. Horaud and Bolles [7] introduced the grouping of features. They extracted features from range images and object models. These features are partitioned in subsets of features with "intrinsic properties". To find the pose of an object in the scene, the features are compared with the help of an ordered tree and a "grow a match" strategy to decrease the processing time. This algorithm is evaluated in an industrial application called 3DPO using local edge-based features. This paper discusses the bin picking process and offers an overview of related work in this decade. Rahardja and Kosaka [8] extract features of complex industrial objects from intensity images. These features are separated into so-called seed features and supporting features. At first, regions are extracted and verified with a priori knowledge (2D appearance of the objects) to get seed features. The depth of the objects in the bin is determined with stereo vision. An iterative process searches the space of corresponding seed features to find the position of the grasp point. Rahardja and Kosaka give an translational error less than 7mm and a rotational error of $10°$ within a processing time of 1.5 minutes on a SUN sparc 1000 server. Hashimoto and Sumi[9] propose an object recognition system which uses range images and intensity images. The process is separated into two parts. At first, information from the range image is used to estimate a rough position of the object. A template of the object shape is matched with the representation of the object in the binarized range image. An intensity image verifies the results of the estimation and refines the position to get an accurate pose with contour matching. The depth of the scene is determined with a structured light stereo vision system. Experiments with box-like objects (600mm x 400mm) have shown that the accuracy of object position has been smaller than 30mm. The recognition process takes about 5 seconds with an unknown computer system. In the paper of Boughorbel et al.[10], an imaging system works with range image and intensity image data to find accurate object positions and to solve the bin picking problem. The geometry of the objects is either reconstructed from the range data or given in form of CAD models. The objects in the scene are modeled with a pre-segmentation and parameter recovery of superquadrics object representation. A camera validates the position and provides additional information for tracking the robot arm. The authors give no information of experimental results. Katsoulas [11] describes a robotic bin picking system in his PhD Thesis. He focuses on a model based object localization system. The model uses geometric parametric entities of deformable superquadrics. The model is compared to scene data, which is preprocessed by edge detection in the input range data. The comparison algorithm performs maximization of the posterior parameters of all known objects to recover the object. Kasoulas proposes a variety of advantages in robustness, accuracy and efficiency. He reported a total processing time of over 100seconds using a Pentium4 at 2.8Ghz for one scene.

3 Preliminaries

3.1 Iterative Closest Points Algorithm

The classical and most commonly used algorithm for rigid transformations is the Iterative Closest Point algorithm [2][12]. The ICP is one of the most popular registration algorithms and was improved by many researchers [1]. Real time implementations [3],[13] show the potential of the ICP Algorithm. As the name already implies, the ICP is an iterative algorithm, that finds a transformation with maximal overlap between two point clouds. In our case, this transformation represents the position of the object present in a scene. The set of points M of the model is defined as:

$$M = \{m_i\}, m_i = \{x_i, y_i, z_i\}, i = 0, 1, 2...N_i \qquad (1)$$

The set of points P of the scene is defined as:

$$P = \{p_j\}, p_j = \{x_j, y_j, z_j\}, j = 0, 1, 2...N_j \qquad (2)$$

The set M consists of points with the coordinates x, y, z in a Cartesian coordinate system. The rigid transformation between the model and the scene is calculated by minimizing the error e

$$e = \sum |m_i - R(p_j) - t|^2 \qquad (3)$$

To find the rotation R and the translation t the closest points between the data sets are determined and used as corresponding point pairs. We use the closed form solution with the help of unit quaternions[14] to find the correct transformation with an point-to-point metric[2]. The algorithm converges monotonously to a local minimum. To stop the iterations there exists mainly the following iteration stop criterions[13]:

- fixed number of iterations
- absolute error threshold
- error difference threshold
- pose change threshold
- complex error criterion

In general all of these iteration stop criterions can be implemented simultaneously. Whenever one of these criterions is met, the iteration stops. As already stated the algorithm minimizes the mean square error of the point distance in several iteration steps, determining the closest points in each iteration step. This leads to a high computational complexity with slow convergence rates and a high sensibility to outliers in the data sets.

3.2   Progressive Meshes

Progressive Mesh is a mesh simplification method focusing upon producing simplified level of detail (LOD) from a mesh of arbitrary size, complexity and shape introduced by Hoppe in 1996[15]. In general the mesh representation consists of the faces (or triangles) defined by three vertices (or points). An edge is the line between two vertices of two adjacent faces. The representation of progressive meshes is given by a set of meshes M_0 to M_n. M_0 is the mesh with the lowest accuracy (the base mesh) and M_n is the mesh with the highest accuracy. The

Progressive Mesh technique represents a sequence of mesh approximations with different levels of details (LOD). Every approximation record includes a base mesh representation and information how to incrementally refine the base representation to the current approximation. The major advantages for our approach are:

– Highly efficient mesh representation
– Level of Detail representation
– mesh simplification with noise reduction

Progressive mesh operations. When dealing with Progressive Meshes, there are two important operations to change the level of detail: *Vertex split* and its opposite operation *Edge collapse.*

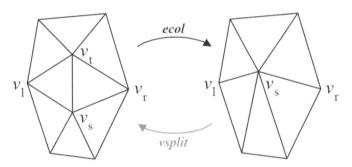

Fig. 1. Progressive Mesh operations

According to Hoppe[15] the edge collapse transformation is show in figure 1. The vertex v_t and v_s are unified into the new vertex v_s. This results in a reduction of the vertices and the number of faces in the mesh for every simplification step. Assuming we have a mesh M with n vertices, this mesh can be simplified by applying an *edge collapse* transformation until the base mesh M_0 is reached. The *vertex split* is the inverted operation to the *edge collapse* transformation. Given the base model M_0, we can add new vertices incrementally to reconstruct the original model. In Figure 1 the vertex v_s is split into two vertices v_t and v_s with the function *vsplit*.

Generating progressive meshes. Generating a progressive mesh means to apply edge collapse transformations to the mesh M_n. The process must be done only once. Like shown in Figure 1 the edge collapse process reduces the edge between v_t and v_s to one vertex. In the process of mesh generation the decision to collapse an edge or not is made with the calculation of an energy function which takes distance of vertices, their number and a regularization term into consideration[16]. The complete process can be described with the following equation:

$$M_0 \underset{edgecollapse}{\overset{vertexsplit}{\rightleftharpoons}} M_1 \underset{edgecollapse}{\overset{vertexsplit}{\rightleftharpoons}} \dots \underset{edgecollapse}{\overset{vertexsplit}{\rightleftharpoons}} M_n \tag{4}$$

Every single step M_i is restored to recover the mesh of this step. So every mesh M_i with the needed accuracy can be retrieved in a very efficient way. The advantage of the progressive meshes implementation is that the simplification can be done in real-time. The Progressive Meshes provide an implicit hierarchical data set for our object localization.

4   Progressive Mesh Iterative Closest Points

The ICP is a time consuming algorithm. The performance mainly depends on the number of points in the scene and model. The major problem of the ICP algorithm is the low performance processing a huge number of points in scene and model. The propose algorithm is a hierarchical system including Progressive Meshes introduced in the previous chapter in order to speed up the convergence process. This combination reduces the complexity of the nearest neighbor search by comparing only the M_i representations of each mesh and not the whole mesh points to each other. The obvious advantage of multi-resolution ICP is the increased performance as already shown in [13][17] and the robustness against outliers. By reducing the mesh up to M_0 most of the outliers cannot longer affect the result of the distance calculation. Multi resolution mesh registration has been used in only a few papers in the past. There exist different mesh representations with its advantages and disadvantages. An approach for multi-resolution mesh registration is introduced by Jost in [13]. Jost increases the number of points by a fixed factor in one LOD step depending on the number of points in the data sets. The author states a convergence speed enhancement of factor 8 by increasing the number of points with sub-sampling when an error criterion [18] is met. Due to this the number of ICP iterations is varying in each resolution step. One year later Zinsser[19] use a uniform subsampling technique in a hierarchical context. They use only every 2^h-th points in the data sets, where h is increased after the data points are aligned with the ICP algorithm. This is combined with robust outlier thresholding and an extrapolation of motion parameters in their *Picky ICP*. Ikemoto et. al[20] aligned warped range data in their coarse-to-fine hierarchical approach. The data sets are separated into small pieces to compensate the global warp, which comes from sensor errors. Each piece is rigidly aligned with respect to other pieces in the data set leading to a better convergence. To gain the advantage of better convergence and accuracy, the total performance is reduced. Their registration tests results in a pairwise match with 1200 points with a total convergence of several minutes using a Pentium4 at 2.8Ghz. The Progressive Mesh representation in combination with the ICP algorithm is introduced by [21]. They register each LOD representation of the mesh with a full ICP registration step in the following way: They create a list of mesh pairs (model and scene) with different LOD's. Each pair is registered with the help of the ICP algorithm starting with the lowest LOD. The resulting transformation of each LOD registration is used as the initial transformation for the next LOD registration. In their experiments they use two LODs for a data set. They report a slightly better performance than the original ICP algorithm (about 5% in convergence time, without considering the time to create the mesh representation). Unfortunately, they do not give any information about the error stop criterion they used. Low [22] smooth their data sets into multiple resolutions by two-dimensional cubic uniform B-spline wavelet transform. They use a pseudo-point-to-plane minimization metric to implement a closed form solution for transformation determination. To change the level of detail, the average distance between points in the current resolution is compared to the average distance between matched points. When the maximum distance between correspondence pairs is less than a 25% the size of the support of the cubic B-spline basis function used to produce the current smoothed surface, the next level of detail is used. Except Low[22] all approaches determine one LOD and try to find the best transformation for exact this model and scene representations with the help of the ICP algorithm. After the best transformation is found with the ICP-Algorithm the LOD is increased and the next ICP process is started in order to find the best transformation for the next LOD. In opposite of this we increase the LOD inside the ICP iteration steps. We reduce the total number of iterations in the registration process over all LODs, which leads to a low convergence time. But the profoundly effect is the increased robustness against outliers[23]. By reducing the mesh (up to M_0) outliers can not longer affect the result of the distance calculation. The shape of the model in representation of M_0 is similar to the M_0 representation of the scene representation to find the initial position in the iterative process of the closest point algorithm. The figure 2 shows the integration of the Progressive Mesh data representation in the iteration steps of the ICP algorithm. Before the iteration starts the model (M_{i1}) and the scene (M_{i2}) data sets are transformed to a coarse representation with a low level of detail. The index i represents the current iteration step and the index starts with (M_{start1} and M_{start2}). The number F of vertices is not necessarily the minimum M_0 of the mesh. According to the experiments it turn out that the optimal number of faces is given by

$$5 \leq F_{start} \leq 0.05 \times F(M_n) \tag{5}$$

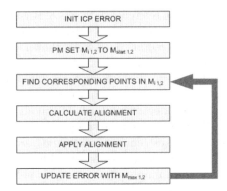

Fig. 2. Progressive Mesh based ICP iteration steps

So the maximum faces at the start of the iteration are five percent of the maximum faces of the mesh. For only one face in each data representation misalignments could occur because the triangle could flip over (face normals are in opposite). Due to this the minimum of 5 faces is used in the proposed algorithm basing on experimental results. The next step is the search of corresponding points in the current LOD data representation M_i. To evaluate our approach against the original ICP algorithm we had to make some assumptions. Because of the fact, that we know the best transformation we defined the ICP iteration stop criterion in our experiment as follows:

$$e_{stop} \leq 1.1 \times min(e_{overall}) \tag{6}$$

We stop the iteration process if the error is below a threshold of 10% over the total minimal error. This criterion can not be used in case the best transformation is not known before. Additionally we use the minimal error difference iteration stop criterion:

$$e_{stop} \leq \tau \; with \; \tau = \sum_{j=1, k=0}^{\eta_{max}} e_{j-k} \tag{7}$$

If the error stagnates over a defined number of iteration steps η_{max}, the iteration is stopped. This criterion let the original ICP converge to a local minimum. The order of the criterions can be set by their parametrization. The main drawback is the loss of performance if every criterion is used. We additionally use another fast criterion: A fixed number of iterations is set to a very high value, so this criterion is met extremely rarely. To overcome a slow convergence over the whole iteration process we stop the iteration after a maximum number of iterations (for example 300 iterations). This stop criterion is met very rarely, because the other criterions are more significant. This combination of iteration stop criterions ensures a quasideterministic convergence time over the process of object localization.

5   Experiments

To evaluate our idea we used the reference data sets of Rusinkiewicz and Levoy[3]. These synthetic meshes have about 100000 vertices added with Gaussian noise and outliers. In Figure 3 the "fractal" data set is shown. The "fractal" data set represents landscape data of terrain registration and has features in all level of details, so this is a good reference for our experiment setup. This test scenario does not cover all possible classes of object surfaces but gives us the opportunity to compare the results with different kinds of ICP modifications like [3] did in their experiments. In [23] we made more test scenarios to show the potential of our approach. We used these test environments and we use no further proven improvements of ICP like selecting, matching, weighting and rejecting points or usage of different metrics and minimizations, to avoid side effects and show the pure results of our contribution. Additionally we calculate the distance error not for the current LOD representation: The sum of squared distances of closest points over all input vertices is calculated in every iteration step, to use this as a

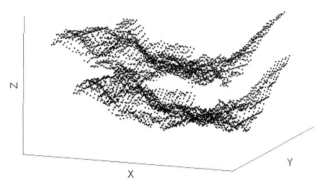

Fig. 3. initial alignment of "fractal" data sets

reference ground truth error. To compare our results we implemented the standard ICP algorithm according to [2] without approximation. We implemented the Progressive Mesh based Iterative Closest Point Algorithm (PMICP) based on the Progressive Mesh implementation of Hoppe in DirectX[24]. All tests are running on a 1,73Ghz Pentium M with ATI Radeon graphic chipset (DirectX 9.0c Support). The convergence theorem of the standard ICP[2] also applies to the PMICP given by the "internal" distance error of the current point data set M_i, but not to ground truth distance error. So our ground truth error can grow in the iterations, but not the distance error of the M_i data set. The squared distance error can not be zero because Gaussian noise is added to the model and scene data set independently. We measure the distance error of the algorithm until it reaches the maximum number of 100 overall ICP iterations. Our experiments show two results: First, the quality of alignment depends on the number of points used to calculate the transformation in each iteration step. On the other hand, if the maximum number of points in the mesh (maximum LOD) is reached, the PMICP converge to the original ICP algorithm. The number of overall iterations is increased, so the process can not converge to the global minimum with an increment size over 25. The optimal LOD step width is around the value 5 with was shown in our experiments[23]. So we increment every iteration step the LOD of our Progressive Meshes (Scene and Model) with this value to achieve the best total convergence time in our experiment. We add Gaussian noise to the fractal scene and measure the convergence time compared to the level of noise. Over all experiments we obtain a 2-3 times lower convergence time for our Progressive Mesh ICP algorithm than the original ICP algorithm. The convergence time of the ICP depends mainly on the number of iterations and nearest neighbor calculation time as already shown in [13]. It can be shown in the experiment data in table 1 that we reduce the number of iterations of the ICP. This can be achieved in addition to the higher computational performance because of the reduced number of points in the nearest neighbor calculation step.

The results of our experiments in table 1 shows that the incremental Progressive Mesh ICP in column 3 is slower than the original algorithm. This is caused by overhead of changing the resolution of the mesh back to the lowest

Table 1. SNR convergence time results

SNR	Original ICP	Inc. PMeshICP	PMeshICP
5	1027.7ms	2623.1ms	335.8ms
2	861.8ms	3099.7ms	371.7ms
1	1238.8ms	4933.1ms	65.2ms
0.5	1140.5ms	3035.3ms	419.0ms
0.2	654.7ms	1149.4ms	115.1ms
0.1	916.2ms	2497.4ms	214.6ms

LOD step in every iteration step. We use the error difference iteration stop criterion and the maximum iterations iteration stop criterion described above for their algorithm. We had made no further improvements to the incremental Progressive Mesh approach, like described in [21]. The complexity of the ICP algorithm depends mainly on the number of points in the data set. The search of the closest points has a computational complexity of $O(N_j \times N_i)$. We reduce the number of points in the data set, starting with only a few points and increase the number in every iteration step. The computational complexity is in average is still $O((N_j) \times (N_i))$ assuming we do not stop the iteration until we reach the end (M_n mesh). If the iteration process is stopped, because the ICP reached the minimum, the performance of our implementation is always four times better than $O((N_j) \times (N_i))$ becuse we use only half of the points in average. Adding the overhead of LOD adjustment in every iteration step, our approach (column 4) is always more than 3 times faster than the original ICP, which is already shown in our convergence time experiments (table 1). Independent to this results the closest point search in our PMICP algorithm can be accelerated by kd-Tree implementation[25] with a complexity of $O((0.5 \times log(N_j)) \times (0.5 \times log(N_i)))$ or even better[26].

6   System Integration

Robotic automation processes became very successful in the last years, and offers a wide range for research. Our approach is used in industrial fields like de-palletizing or robotic bin picking. The last chapters focused on the object localization step, which is the most challenging step in the whole process. Object recognition and object localization has a long history in two dimensional image processing[27], but using distance data provided by laser range sensors our approach is able to find objects in three dimensional (3D) scenes. Many of the known solutions for bin picking are limited to simple shaped objects or objects with specific features. But in many industrial automation processes the handling of objects with complex shapes without any specific features is still an unsolved problem in the field of robotic automation[11]. There exist many advantages to use range data provided by 3D laser range sensors. The industrial robot use the coordinates of the found object to pick up the object. We set up a system with an industrial robot (Kuka KR6) and an industrial laser range sensor (Sick LMS400)like shown in figure 4. In this application a KUKA robot (Kuka KR6 Robot, 2007) is used to pick up the objects. It is a six-axis industrial robot with a special gripper for this brake disks. The robot KUKA KR6 is characterized by load limit of 6kg and a repeatability of 0.1millimetres. The robot controller is connected to a PC and receives the object positions via serial communication port. The sensor is mounted on a conveyor belt moving across the bin with known objects. The LMS400 sends a laser beam towards a rotating mirror to scan a two-dimensional profile sending out and receiving laser beams in an angle of 70°. The laser measuring system LMS400 is based on the LMS200 and was developed for close range scans up to 3m. The LMS400 has a distance resolution of 1mm at a maximum measuring distance of 3 meters. The typical angular resolution at 250 Hz is 0.1°. Lower sample frequencies slow down the measurement process. The measurement range starts with at 700mm (SICK AG, 2007). The lowest angular resolution of 0.1° yields to an area of 700 distance values at a maximum usage of the angular range of 70°. The measured sensor distance data is provided to a PC via Ethernet. Additional remission values are delivered. The sensor records distance values in one line and transfers the data of one line (scan) to a PC. After each line the sensor is moved a little step further to take the next distance values. Because of the trigger signal from the step motor to the laser sensor, the scans can be recorded in equal distance steps. The picture 4 shows brake disks lying in a bin. The distance values are represented as grey values in the picture. Darker pixels represent closer distance values, objects

Fig. 4. setup of our robotic bin picking system

with brighter pixel are far away. Most of the brake disks are arranged planar, so the object localization is an easy task for feature based localization. To find the hole in the middle of each brake disk, this chapter describes a fast and simple algorithm. The first part of the algorithm bases on commonly used 2D image processing methods. Therefore, two circles with a fixed radius must be found representing the topmost ring of each brake disk. There exist many different methods like contour based matching or Generalized Hough transform (GHT) to find the known pattern in an image. We use an image processing tool like described in [28] to find the best matching brake disk in the image taking orientation, scale and occlusion into consideration. The matching algorithm needs the number of objects to be found and the template in form of the manual created circles. The search can be minimized because of the rotational invariance and the given maximum slope of the objects in this application. The library delivers the best matching templates in the image and information about their similarity, scale and slope referring to the template. Additionally the coverage is determined so occlusions can be found very easy. All these results are taken in consideration for one quality value which calculated for each found brake disk. Around the center point of the chosen disk four small regions are used to find the orientation of that disk. The regions are shifted by 90 degrees to each other. Inside these rectangular regions the median grey values are extracted. Each grey value is assigned to a specific distance value. The distance to the sensor can be recalculated with the help of this grey value. The cross product of this vectors delivers the orientation of the plane, given by the four points representing the brake disk. With the known depth d of the brake disk the object reference point in the middle of the brake disk can be calculated. This is shown in figure5. As

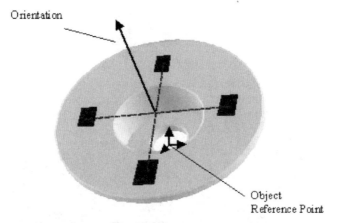

Fig. 5. orientation determination of a brake disk

stated in section 4 we need scene and model points for the alignment. To get model points we simulate the LMS400 industrial laser range sensor[29]. The pose estimation transfers a Computer-aided design (CAD) model to a point representation taking the properties of the simulated laser range sensor in consideration.

The model representation is transformed to coarse pose of the candidate of the pose estimation[30]. These two data sets are aligned with the Progressive Mesh based ICP in the pose refinement step 4. The determined transformation vector of each simulated object is transformed to the global coordinate system and transferred to an industrial robot to pick up the object. Because of the high performance implementation of the library, the algorithm needs not more than 1 second to find the best object candidate. Further test scenarios the algorithm proofs his robustness against brake disks with extreme orientation up to 50°. We placed brake disks manually in steps of 10° of orientation in the bin. The algorithm delivers the position within ±2-3 millimeters and ±1-2° in orientation. Additionally the robot itself has also an absolute accuracy of 1-2 millimeters. This accuracy is sufficient in this application, because the gripper is able to adjust the position in this range.

7 Conclusion

We described a fast object localization system to align range data surfaces with an Progressive Mesh based multi-resolution ICP algorithm in a context of industrial process automation. The key idea of this paper was that we change the level of detail of our model scene data sets within the ICP iteration steps using a high efficient data representation: Progressive meshes. We focus on the improvements in performance and robustness and compare our contribution to the basis algorithm and similar approaches. To be sure to meet the requirements of different applications we evaluated our system with test scenarios and a real scenario for robotic bin picking. One of the promising improvements in the near future will be the modification of the ICP algorithm using face normals for point-pair rejection. Many of already known modification can be applied to our approach. Using an automatic level of detail adjustment will further increase the robustness and reduce the computational costs. So this offers many possible extensions.

Acknowledgement

The authors would like to thank S. Rusinkiewicz for supplying the test scenario data sets. The author would also like to thank the following persons for their support: M. Otesteanu, P. Roebrock, M. Kleinkes, W. Neddermeyer, W. Winkler, and K. Lehmann. This project has been partly funded by VMT GmbH(Pepperl+Fuchs Group).

References

1. Salvi, J., Matabosch, C., Fofi, D., Forest, J.: A review of recent range image registration methods with accuracy evaluation. Image Vision Comput. 25(5), 578–596 (2007)
2. Besl, P., Jain, R.: Three-dimensional object recognition. Computing Surveys 17, 75–145 (1985)

3. Rusinkiewicz, S., Levoy, M.: Efficient variants of the icp algorithm. In: Levoy, M. (ed.) Proc. Third International Conference on 3-D Digital Imaging and Modeling, pp. 145–152 (2001)
4. Brady, J., Nandhakumar, N., Aggarwal, J.: Recent progress in the recognition of objects from range data. In: Proc. International Conference on Pattern Recognition, November 14-17, pp. 85–92 (1988)
5. Kak, A.C., Edwards, J.L.: Experimental state of the art in 3d object recognition and localization using range data. In: Proceedings of the IEEE Workshop on Vision for Robots (1997)
6. Ikeuchi, K., Horn, B.K.P., Nagata, S., Callahan, T., Feingold, O.: Picking up an object from a pile of objects. In: MIT AI Memo, pp. 139–166 (1983)
7. Bolles, R.C., Horaud, P.: 3dpo: A three-dimensional part orientation system. Int. J. Rob. Res. 5(3), 3–26 (1986)
8. Rahardja, K., Kosaka, A.: Vision-based bin-picking: Recognition and localization of multiple complex objects using simple visual cues. In: Proceedings of IEEE/RSJ International Conference on Intelligent Robots and Systems, vol. 3, pp. 1448–1457 (1996)
9. Hashimoto, M., Sumi, K.: 3-d object recognition based on integration of range image and gray-scale image. In: BMVC (2001)
10. Boughorbel, F., Zhang, Y., Kang, S., Chidambaram, U., Abidi, B., Koschan, A., Abidi, M.: Laser ranging and video imaging for bin picking. Assembly Automation 1, 53–59 (2003)
11. Katsoulas, D.: Robust Recovery of Piled Box-Like Objects in Range Images. PhD thesis, Albert-Ludwigs-Universität Freiburg (2004)
12. Chen, Y., Chen, Y., Medioni, G.: Object modeling by registration of multiple range images. In: Medioni, G. (ed.) Proc. IEEE International Conference on Robotics and Automation, vol. 3, pp. 2724–2729 (1991)
13. Jost, T., Hugli, H.: A multi-resolution scheme icp algorithm for fast shape registration. In: Proc. First International Symposium on 3D Data Processing Visualization and Transmission, June 19-21, pp. 540–543 (2002)
14. Horn, B.K.: Closed-form solution of absolute orientation using unit quaternions. Journal of the Optical Society of America A 4(4), 629–642 (1987)
15. Hoppe, H.: Progressive meshes. In: SIGGRAPH 1996: Proceedings of the 23rd Annual Conference on Computer Graphics and Interactive Techniques, pp. 99–108 (1996)
16. Hoppe, H., DeRose, T., Duchamp, T., McDonald, J., Stuetzle, W.: Mesh optimization. In: SIGGRAPH 1993: Proceedings of the 20th Annual Conference on Computer Graphics and Interactive Techniques, pp. 19–26. ACM, New York (1993)
17. Zhang, Z.: Iterative point matching for registration of free-form curves and surfaces. Int. J. Comput. Vision 13(2), 119–152 (1994)
18. Schuetz, C.: Geometric point matching of free-form 3D objects. PhD thesis, University of Neuchâtel (1998)
19. Zinsser, T., Schmidt, J., Niemann, H.: Performance analysis of nearest neighbor algorithms for icp registration of 3-d point sets. In: Ertl, T., Girod, B., Greiner, G., Niemann, H., Seidel, H.P., Steinbach, E., Westermann, R. (eds.) Vision, Modeling, and Visualization 2003, Munich, Germany, pp. 199–206. Aka/IOS Press, Berlin (2003)
20. Ikemoto, L., Gelfand, N., Levoy, M.: A hierarchical method for aligning warped meshes. In: Proc. 4th International Conference on 3D Digital Imaging and Modeling (3DIM), pp. 434–441 (2003)

21. Chen, X., Chen, X., Zhang, L., Tong, R., Dong, J.: Multi-resolution-based mesh registration. In: Zhang, L. (ed.) Proc. 8th International Conference on Computer Supported Cooperative Work in Design, vol. 1, pp. 88–93 (2004)
22. Low, K.L., Lastra, A.: Reliable and rapidly-converging icp algorithm using multiresolution smoothing. In: Fourth International Conference on 3-D Digital Imaging and Modeling (3DIM 2003), p. 171 (2003)
23. Boehnke, K., Otesteanu, M.: Progressive mesh based iterative closest points for robotic bin picking. In: Proceedings of the International Conference on Informatics in Control, Automation and Robotics, pp. 469–473 (2008)
24. Hoppe, H.: Efficient implementation of progressive meshes. Computers & Graphics 22(1), 27–36 (1998)
25. Simon, D.: Fast and Accurate Shape-Based Registration. PhD thesis, Robotics Institute, Carnegie Mellon University, Pittsburgh, PA (December 1996)
26. Greenspan, M., Yurick, M.: Approximate k-d tree search for efficient icp. In: Fourth International Conference on 3-D Digital Imaging and Modeling (3DIM 2003), p. 442 (2003)
27. Andrade-Cetto, J., Kak, A.C.: Object Recognition. Wiley Encyclopedia of Electrical and Electronics Engineering (2000)
28. Boehnke, K., Otesteanu, M., Roebrock, P., Winkler, W., Neddermeyer, W.: An industrial laser scanning system for robotic bin picking. In: 8th Conference on Optical 3-D Measurement Techniques (2007)
29. Boehnke, K.: Object localization in range data for robotic bin picking. In: Proc. IEEE International Conference on Automation Science and Engineering CASE 2007. (2007) 572–577
30. Boehnke, K.: Fast object localization with real time 3d laser range sensor simulation. WSEAS Transactions on Electronics: Real Time Applications with 3D Sensors 5(3), 83–92 (2008)

What Role for Emotions in Cooperating Robots?
– The Case of RH3-Y

Jean-Daniel Dessimoz and Pierre-François Gauthey

HESSO-Western Switzerland University of Applied Sciences,
HEIG-VD, School of Management and Engineering,
CH-1400 Yverdon-les-Bains, Switzerland
{Jean-Daniel.Dessimoz,Pierre-Francois.Gauthey}@heig-vd.ch

Abstract. The paper reviews key aspects of emotions in the context of cooperating robots (mostly, robots cooperating with humans), and gives numerous concrete examples from RH-Y robots. Emotions have been first systematically studied in relation to human expressions, and then the shift has come towards a machine-based replication. Emotions appear to result from changes, from convergence or deviation between status and goals; they trigger appropriate activities, are commonly represented in 2D or 3D affect space, and can be made visible by facial expressions. While specific devices are sometimes created, emotive expressions seem to be conveniently rendered by a set of facial images or more simply by some icons; they can also possibly be parameterized in a few dimensions for continuous modulation. In fact however, internal forces for activities and changes may be expressed in many ways other than faces: screens, panels, and operational behaviors. Relying on emotions ensures useful aspects, such as experience reuse, legibility or communication. But it also includes limits such as due to the nature of robots, of interactive media, and even of the very domain of emotions. For our goal, the design of effective and efficient, cooperating robots, in domestic applications, communication and interaction play key roles; best practices become evident after experimental verification; and our experience gained so far, over 10 years and more, points at a variety of successful strategic attitudes and expression modes, much beyond classic human emotions and facial or iconic images.

Keywords: Emotions, cooperating robot, Robocup-at-home, domestic help, cognitics.

1 Introduction

The world keeps changing, and so do robots, humans, and robots cooperating with humans. Now what makes humans change, and drives their actions: emotions. Do we need similar driving forces for robots? Should those be similar to human emotions? Furthermore do we need somehow to express emotions, and recognize them in others, i.e. to *communicate* emotions? Here are the topics addressed by this paper, along with concrete examples from RH3-Y, a cooperating robot designed for helping at Home.

A. Gottscheber, D. Obdržálek, and C. Schmidt (Eds.): EUROBOT 2009, CCIS 82, pp. 38–46, 2010.

Emotions and their expressions have started to be systematically explored more than 30 ears ago [1], and keep receiving a lot of attention (e.g. [2]). More than fifty different kinds of emotions have been identified (e.g. [3]). They have been studied by psychologists, and found very much expressed on the face of humans (e.g. [4]). Nowadays, emotions and their expressions are even viewed as multidimensional continua, in spaces of dimensions 2 (e.g. [5]), 3 (e.g. [6]), or even more (e.g. 2x3 mapping from emotion basis to expression complex – eyes, lips and eyebrows [7]) Innovative expressions have been attempted in specific physical context (e.g. Kismet ears [6], or Probo trunk [8]), or in virtual world (e.g. [5]). We find that iconic representations are more easily expressed, and also understood, than actual human faces; and that functional behaviors deliver often more direct and clearer messages than conventional or cultural attitudes. This can easily be illustrated in a case study, relating to RH3-Y robot [9]. RH3-Y was designed as an autonomous, cooperating robot, as for domestic applications in Robocup-at-Home league (RAH) [10], and features numerous cognitic capabilities [11].

The paper is organized as follows. Part 2 defines emotions and presents basic models for them and their expression. Part 3 presents images and icons as mutual alternatives for emotional expressions. Part 4 reviews other ways to express emotions, with many specific examples taken form the case of RH3-Y. Part 5 summarizes considerations about the role of emotions, and of their expression, for a cooperating robot.

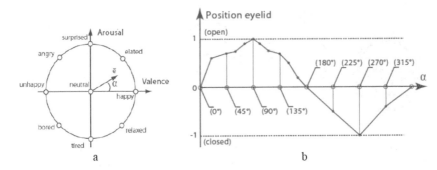

Fig. 1. Eight discrete emotions are cyclically represented in 2D, arousal-valence space (a). Corresponding eyelid positions are tabulated for the case of Probo [8], in 10 to 20 values (b).

2   Models for Emotions

Change is general and often very abstract. It could be defined as the inverse of time. Physical analogies are often used in natural language in order to describe more abstract concepts, and in that way "changes" are often described in reference to locations and to space dimensions: motion, speed, acceleration, forces, stability, etc.

Emotions, by their etymology, clearly relate to motion. Resulting from changes in environment, perception and possibly projected consequences, related to convergence or deviation between status and goals, emotions can be considered as psychological forces that trigger subject activities towards strategic goals.

Experts suggest representing emotions as vectors in a 2D or a 3D mathematical space with the following primitives:

- Arousal. Arousal denotes an activity level. Associated aspects include the ones of energy, quantity of perceived information, urgency and intensity of desired changes, of planned actions. Arousal is always positive (or zero).
- Valence. Valence denotes a happiness degree; it can be either positive or negative. It might be interpreted as the current balance, as subjectively perceived, of overall "benefits" and "costs"
- Stance. Taken with less priority into account, stance is an attitude that may vary between "open", open to dialogue, to empathy or on the contrary "closed", barring exchanges and cooperation.

As an example, in the fig.1 below, from [8], eight discrete emotions are cyclically represented in 2D space, with a predefined polar mapping onto eyelid positions for corresponding expressions.

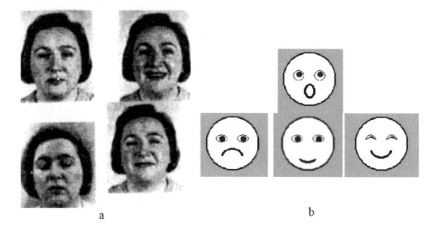

a   b

Fig. 2. Facial expression of emotions in positive valence and low arousal quadrant, from [4] (a). And icons in middle to high arousal, negative to positive happiness valence (b).

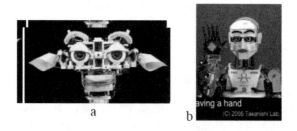

a   b

Fig. 3. With many mobile parts, "Kismet" has been designed for the expression of emotions, for a machine exploring interaction with humans [6] (a); on the right, the robot WE-4RII of Takanishi lab, Waseda University is also a humanoid capable of multiple emotional expressions.

Emotions have been defined above as they typically apply in human psychology. We were relying there on physical analogies (forces, motion, etc.). Now if we transpose to machines, analogies may no longer be necessary (unless of course one wants explicitly to refer to emotions). Here is the phrase from above, rewritten in a style closer to standard engineering: Resulting from changes in environment, perception and possibly projected consequences, computed in particular from errors between set points and measurements, robot behavioral decisions can be considered as internal commands that trigger actions toward strategic goals.

Even though we recognize here the classic paradigm of closed loop control, and even though this might be appropriate internally in machines, in the context of this paper and of cooperating robots however, emotions cannot be escaped, at least for their components being mutually expressed and observed between robots and humans. For interaction between robots and other robots, totally novel approaches can be envisioned, and the concept of emotions is not a clearly interesting approach.

Emotions are interesting concepts here, as they may be communicated (shared, exchanged, in the classical terms of information) between humans and robots. There are many other concepts yet, inspired by biological systems, such as e.g. neurons, genetic algorithms, believes, or intentionality, which, in our opinion, are less important; first, the latter are not as obviously apparent in human-robot communication as emotions often are, and second (this however is not discussed in this paper), they usually do not ensure better performances as alternatives in classical engineering.

3   Images and Icons

In general, the communication of emotions relies primarily on facial expressions.

Although naturally expressed in full (3D) space, most of emotional content can also be perceived on the basis of simple images.

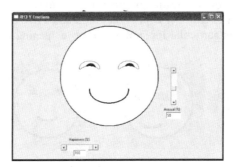

Fig. 4. The expression of emotions can be parameterized and continuously adjusted in terms of mouth shape and eye lid location [12]

It turns out that yet much simpler representations, such as caricatures or icons, can not only retain essential emotion related messages but also may even be more expressive. Looking at Fig. 2., where some images of real people are presented along with a few RH3-Y samples of iconic facial expressions, can easily qualitatively validate this.

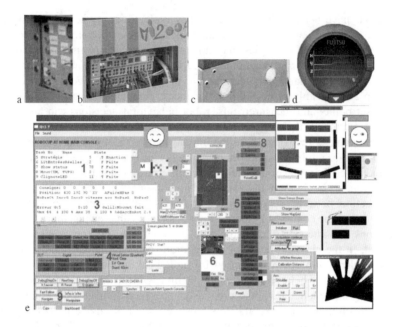

Fig. 5. The expression of status and intentions are classically described by panels, displays and screens (a: Hornuss, b: Dude, c: programmable "eyes" or modulated headlights, RH3-Y; d: 3-D acceleration components on supervising computer; e: set of RH3-Y interactive control screens).

Sometimes people attempt to express machine emotions by a real, 3D, dedicated physical structure. Fig.3 provides an example of such solution. Even though such approaches have some advantages, such as consisting itself as a display medium, the efficiency in terms of communication of emotions is – to say the least -not obvious.

Extending the idea of icons, it is possible to parameterize expressions so as to get continuous changes, in particular for smiles or for eye opening (fig.4).

Fig. 6. The expression of robot status could evolve towards special adaptations of icons (a: "mute" behavior; b: leftwards intentionality; c: idem upwards; d: acknowledgement of common understanding.

4   Other Ways to Express Emotions

It has been mentioned that emotions, i.e. internal ("psychological") forces for behavioral changes, are primarily expressed by specific facial patterns.

Classically however, robots such as ours can also show their status, processes and intentions, i.e. more or less explicitly rendering their "emotions", on various control screens, devices and panels.

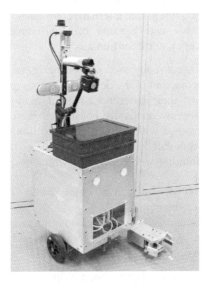

Fig. 7. General view of RH4-Y robot

Now these classical data could be more systematically linked with iconic parameterizations (re. Fig. 6), leading to new head and facial expressions.

RH-Y can express its emotions in many other ways yet: blinking (since 1998, considering previous developments in Eurobot context), speaking, delivering predefined sound waves (re. RH2-Y in Atlanta), or moving its body and its arm in a way quite similar to dancing.

And most naturally, those who know RH3-Y also recognize some of its emotive states, such as when its motors are servoed, or how its right wheel tends to lag behind, when batteries run tired.

5 What Role for Emotions in Cooperating Robots

Emotions help explaining, in psychology, how humans make their short-term strategic choices. Emotions set a person in motion into a certain direction, towards a certain goal.

Now does referring to emotions make sense for robots? For some purposes it is useful, and for others it is not.

Useful aspects. The useful aspects of dealing with emotions include experience, legibility and communication.

Humans inherit thousands of years of experience, relating to emotions. Attempting to transfer even just part of it to machines could be worth the effort.

Describing machine attitudes and behaviors in terms of emotions may get them familiar and immediately understood by humans for design as well as for developmental, operational, and debugging purposes.

Improved communication may be the strongest advantage provided by emotional messages: thus in addition or as an alternative to other means, the exchange of information between robots and humans can be conveyed through a channel (emotional expressions) very natural for humans.

Limits. On the other hand, emotion-based approaches have also strong limits due to the nature of robots, of media, and of the domain of emotions itself.

Robots are not humans; their differences are far more numerous than potential similarities. Therefore it is now generally considered preferable to let users clearly recognize that they do *not* deal with humans, so as to limit impossible expectations. Furthermore, the same argument of legibility plays here the opposite role for those observers who are more familiar with concrete machine peculiarities, rather than with human affect psychology.

Robots feature different media, such as color screens or blinking lights, which allow for novel communication paths.

And in fact even for humans, the domain of emotions is *not* so well known today, and therefore transposing classical knowledge of this field to machines means also transferring current uncertainties.

Synthesis. Let's attempt a balanced conclusion, drawing from general considerations as well as from specific Robocup at Home (RAH), and RH3-Y contexts. Here communication is also a key function to consider.

It is universally true that reality is minimally upon reach. Therefore it is critical here also to set the priority on a goal oriented strategy.

Our goal relates to robotic help at home. And the latter has been further specified in RAH tests and rulebook.

Cooperation between humans and robots is very important in this context. Communication helps in synchronization, e.g. asking for or providing mutual help, coordinating intentionality, interest, attention, vergence, possibly threatening. For communication purpose, and more generally successful interaction, expression seems mandatory. Very often however the latter simply result from functionality. E.g. an emotion may induce a backward robot gesture; and for the observer just perceiving that gesture may be sufficient to get aware of inducing emotion.

Numerous examples for communicating emotions, in the study case of RH3-Y, have been given above. Another interesting one (the latest one implemented and experimented with RH-Y) is the acknowledgement of user commands in switching active or resting modes for the "FastFollow" test: within a fraction of a second, the robot can react with a green or red panel color and diodes to control gestures of human users. This provides a significant improvement with respect to previous situations where dialogue was based on the perception of potential changes in robot motions (about ten times faster reactions).

Another point relates to the nature of machines, which being different from humans, gains in being granted, beyond classic human emotions, a wealth of variety in terms of internal affect forces and strategic attitudes.

6 Conclusion

The paper has concisely reviewed key aspects of emotions, in the context of cooperating robots, with numerous concrete examples taken from RH-Y robots.

In the past, emotions have first been systematically studied in relations to human expression, and then the shift has come towards machine-based replication.

Emotions appear to result from changes, from convergence or deviation between status and goals, and to trigger appropriate activities. They are commonly represented in 2D or 3D affect space, and made visible by facial expressions.

While specific devices are sometimes created in ad hoc way, emotive expressions seem also to be conveniently rendered by a set of facial images or more simply icons; possibly parameterized in a few dimensions, for continuous modulation.

In fact however, internal forces towards activities and changes may be expressed in many ways other than just faces: typically, screens, panels, and operational behaviors may contribute to sharing emotions.

Relying on emotions ensures useful aspects, such as experience reuse, legibility or communication. It also includes limits however, such as due to the nature of robots, of media, and even of the very domain of emotions. For our goal, which is the design of effective and efficient, cooperating robots, in domestic applications, the ability to communicate and to interact plays a key role. Best practices become evident after experimental verification, and tend to point at a variety of strategic attitudes and expression modes, much beyond classic human emotions and facial or iconic images.

In fact our experience shows, after many years of operation with autonomous robots, that communication, while strongly asymmetrical, works rather well between robots and humans. In the direction from humans to robots, communication relies on engineered paths and software based strategies; in the direction from robots to humans, in addition to resources similar to these, the mere functional behavior of robots convey a wealth of effective information to humans, especially when the latter are deeply accustomed to their operation and even involved in their design.

References

1. Ekman, P., Friesen, W.: Facial Action Coding System - A Technique for the Measurement of Facial Movement. Consulting Psychologists Press, Palo Alto (1978)
2. National Centre of Competence in Research (NCCR) for the Affective Sciences, Geneva, Switzerland: Emotions (March 2009), http://www.affective-sciences.org/emotion-details
3. Wikipedia Foundation: More than 50 emotions listed (March 2009), http://en.wikipedia.org/wiki/Emotion
4. Russell, J.A.: Reading emotions from and into faces - Resurrecting a dimensional contextual perspective. In: Russell, J.A., Ferna_andez-Dolz, J.M. (eds.) The Psychology of Facial Expression, pp. 295–320. Cambridge University Press, Cambridge (1997)
5. Garcia-Rojas, A., Vexo, F., Thalmonn, D., Roouzaiou, A.: Emotional face expression profiles supported by virtual human ontology, Comp. Animation Virtual Worlds 2006, vol. 17, pp. 259–269. John Wiley & Sons, Ltd., Chichester (2006)
6. Breazeal, C.L.: Sociable Machines: Expressive Social Exchange Between Humans and Robots, Dr Sc. Thesis, Massachusetts Institute of Technology, 20 May (2000)

7. Lim, M.Y., Aylett, R.: A New Approach to Emotion Generation and Expression, DC. In: The 2nd International Conference on Affective Computing and Intelligent Interaction, Lisbon, (September 12-14, 2007)
8. Goris, K., Saldien, J., Lefeber, D.: The Huggable Robot Probo, a Multi-disciplinary Research Platform. In: Proc. Eurobot Conference 2008, Heidelberg, Germany, May 22-24 (2008) ISBN: 978-80-7378-042-5
9. Dessimoz, J.-D., Gauthey, P.-F.: RH3-Y – Toward A Cooperating Robot for Home Applications, Robocup-at-Home League. In: Proceedings Robocup08 Symposium and World Competition, Suzhou, China, July 14-20 (2008)
10. RobocupAtHome League (2007), http://www.robocupathome.org
11. Dessimoz, J.-D., Gauthey, P.-F.: Quantitative Cognitics and Agility Requirements in the Design of Cooperating Autonomous Robots. In: Proc. Eurobot Conference 2008, Heidelberg, Germany, May 22-24 (2008) ISBN: 978-80-7378-042-5
12. Dessimoz, J.-D.: West Switzerland University of Applied Sciences. website for HEIG-VD RH4-Y cooperating robot for Robocup-at-Home context (March 2009), http://rahe.populus.org/rub/3

A Mobile Robot for Small Object Handling

Ondřej Fišer, Hana Szűcsová, Vladimír Grimmer, Jan Popelka,
Vojtěch Vonásek, Tomáš Krajník, and Jan Chudoba

The Gerstner Laboratory for Intelligent Decision Making and Control
Department of Cybernetics, Faculty of Electrical Engineering
Czech Technical University in Prague
{fisero1,szucsh1,grimmv1,popelj3}@fel.cvut.cz,
{vonasek,tkrajnik,chudoba}@labe.felk.cvut.cz

Abstract. The aim of this paper is to present an intelligent autonomous
robot capable of small object manipulation. The design of the robot is
influenced mainly by the rules of EUROBOT 09 competition. In this
challenge, two robots pick up objects scattered on a planar rectangular
playfield and use these elements to build models of Hellenistic temples.
This paper describes the robot hardware, i.e. electro-mechanics of the
drive, chassis and manipulator, as well as the software, i.e. localization,
collision avoidance, motion control and planning algorithms.

1 Introduction

1.1 EUROBOT

The EUROBOT[1] association holds amateur robotics open contests, organized
either in student projects or in independent clubs, or in educational projects. It
originated in 1998 in France as the national robotics cup. In a typical EUROBOT
match, two autonomous mobile robots compete on a planar rectangular field.
Robots are limited in size - the maximum height is 0.35 m and their convex hull
circumference must not exceed 1.2 m. After the start of the match, the robot
can deploy its devices and extend its perimeter up to 1.4 m. The match lasts
90 seconds.

The competition rules change every year. This prevents from the situation in
other leagues (e.g. FIRA[2]), where new teams are disadvantaged compared to
veteran participants and are seldom able to reach a good rank. The challenge of
this year is called "Temples of Atlantis".

In this challenge, two robotic adversaries attempt to build models on Hellenistic temples. In order to do so, the robots have to find and gather construction
material - i.e. column elements and lintels. After that both opponents have to
deliver construction elements to designated building areas. Inside of these areas,
the gathered objects have to be put together in an ancient temple resembling
shape. The number of construction elements carried by the robot is limited and
therefore an elaborate temple model building requires the robot to perform several "runs", each comprising of gathering and building phases. The match is
determined by rating assigned to temples built by each competing robot. Detailed rules description can be found at [3].

A. Gottscheber, D. Obdržálek, and C. Schmidt (Eds.): EUROBOT 2009, CCIS 82, pp. 47–60, 2010.

1.2 Project Goal

The participation in such competition allows good opportunity to compare advanced AI algorithms to simple "state machine" solutions in a competitive scenario. Sensing and planning algorithms designed for these systems have to take into account limited computational resources of the robots. These algorithms often find their utilization in other projects which rely on small-scale robots.

However the major objective of participation in this competition is educational. In recent years, the competition has proved to be an excellent opportunity for a group of students to test their theoretical knowledge and practical skills in a complex project. Participating students have not only solved problems related to different areas of engineering, but also trained their "soft" skills. Moreover, the competition provides a unique chance to compare particular approaches and stimulates knowledge exchange between international teams. This is a major difference compared to a common educational project, when students solve only particular tasks and the solution is often known in advance.

1.3 Paper Structure

The paper is organized as follows. It starts with introduction. Next chapter is concerned with an overview of robot concept and desired activity. This is succeeded by sections describing the robot motion, manipulator, sensing and planning subsystems. The subsequent chapter concludes about robot ability to fulfill the desired task. The conclusion is followed by acknowledgments and references.

2 Overview

2.1 Robot

Our robot hardware consists of a alluminium chassis, differential drive, two mechanical grippers, power subsystem, three micro-controller units (drive, manipulator and odometry), onboard PC, mechanical bumpers, laser and infrared rangefinders. The robot software is divided into localization, planning, motion control and manipulation modules. The onboard computer is equipped with Ångström Linux operating system. Software modules running on a PC are programmed in C++ and control algorithms are written in C.

The skeleton of the robot is formed by interlocked X-shaped alluminium beams. These proved to be firm and stable enough to support robot devices and provide reliable shock protection. The remaining parts of the chassis which support sensors and electronics are made of alluminium, duraluminium or printed circuit boards.

The drive subsystem has its own control unit capable of positioning boths wheels independently. The motion control module realizes a set of reactive behaviours (e.g. docking to the building area, approaching column elements) and a position regulator. Two bumpers and two infrared rangefinders serve as primary

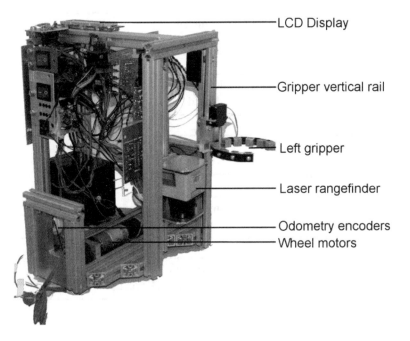

LCD Display

Gripper vertical rail

Left gripper

Laser rangefinder

Odometry encoders
Wheel motors

Fig. 1. Robot overview

collision detection sensors. Aside from signals from these sensors, the motion control board monitors signals from start trigger, emergency stop and other function buttons. Odometry is measured by two passive wheels with optical encoders and an independent microcontroller.

The localization module is based on odometry and laser rangefinder [4] with variable scanning height. The circular building area and walls serve as primary landmarks for localization. Opponent robot position is calculated and taken into account in the planning module.

The manipulator mechanism consists of two vertically movable grippers. The mechanism allows to hold one lintel and two columns. However several construction elements can be placed on the carried lintel and therefore the robot can carry four columns and two lintels simultaneously.

Planning subsystem implements a predefined Petri net [5], where places represent functions of individual software modules and transitions describe return values of these functions. In our approach, a place in a Petri net represents some functionality of a particular robot subsystem (e.g. manipulator: grab columns).

This year, building blocks of a particular team are distributed near the teams' starting point and therefore most of the robots will meet only at the building areas. Opponent robot position should be therefore considered when choosing where to build a temple, but sophisticated collision avoidance does not seem to be necessary. Therefore, we implemented only a simple reactive collision avoidance.

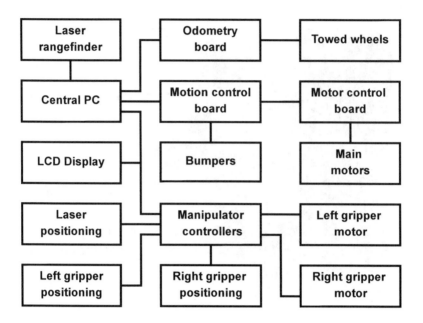

Fig. 2. System scheme

2.2   Match Tactics

A lintel is placed on the robot grippers prior to the match. At the start of each round, the robot scans the playfield and determines the configuration of randomly placed columns. After that, it attempts to grab two column elements. When both columns are picked up, the laser rangefinder is raised in order to acquire the opponent position and speed. Based on this data, the building area is chosen and both grippers adjust their height accordingly. As soon as the building area is approached, both grippers release columns and lower themselves in order to put a lintel on the columns. After the first "level" of a temple model is completed, the laser rangefinder is lowered in order to scan for walls and construction elements. The robot will approach the lintel dispenser and get the lintel on both grippers. In order to build another level, the aforementioned procedure is repeated. Alternatively, the robot attempts to pick up another lintel dispenser immediately after acquiring the first two columns. The robot then attempts to grab additional two columns and builds a large temple at a time.

Most of the time, the laser rangefinder serves as primary localization and collision avoidance sensor. When approaching the building area, the rangefinder position is changed to scan for the opponent and acquire information about the situation on the building site. If the adversary is detected near the construction area we want to build on, our robot will wait if the opponent is going to clear the way. In case the path is not clear after a moment, planning module will choose an alternative building site. When leaving the building site, the robot moves backwards and relies on rear infrared rangefinder and mechanical bumper to detect collisions.

3   Motion

The motion subsystem is in charge of moving the robot on the field while preventing damage done to our or opponent robot. It consists of two wheels with DC motors, motor control board, motion control board, mechanical bumpers and rear infrared rangefinders. We use EGM30 12V/4.5W motors equipped with encoders and a 30:1 reduction gearboxes. The MD23 dual motor driver is able to control either speed or position of both motors independently. It can also report the current state of the IRC counters attached on both motors. Acceleration and speed are monitored and limited by this controller to suppress wheel slipping. This motor controller is connected to the motion control board via an I^{2}C bus. The motion control board handles inputs from bumpers and infrared rangefinders and issues commands to the motor controller.

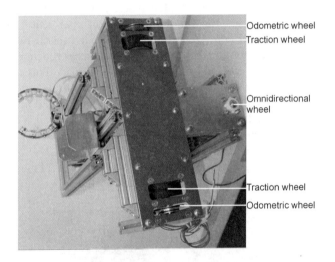

Fig. 3. Robot bottom view

In addition, it processes signals from buttons used during robot starting procedure. These buttons are "system test", "change color", "change strategy" and "start". The "start" button is connected paralely with a start cable connector.

This control board receives orders from onboard PC via a RS232 interface and translates them to commands for the motor controller. The motion control unit stores latest information from bumpers and when requested, it transmits these data to the onboard PC by RS232 interface. Signal "bouncing" during the switch transitions is suppressed by RC filters.

Collision detection, realized by mechanical bumpers prevents damage by a robot attempting to move in a blocked direction. In such situations, rotating wheels erode the field surface - the robot, which damages field gets disqualified. Bumpers are composed of a microswitch covered by a small metal plate.

Two basic operating modes are realized by the motion subsystem. First operating mode consists of a set of reactive behaviors. These translate the rangefinder and bumper signals to commands for the motion board. The second mode realizes a "follow the carrot"[6] regulator. In this mode, the onboard PC sends robot-relative coordinates of a target position and the robot moves towards it.

4   Manipulator

The manipulator serves not only for item pickup and transport, but also for the temple model construction. It consists of two vertically positioned grippers, a control board and additional sensors.

The gripper itself consists of two floppy bands, fulfilling roles of fingers, a DC motor and an absolute rotation sensor. Each floppy band has several metallic cubes attached to one side. A steel cord is reeved through these cubes and attached to the cube at the band end. In the gripper center, a DC motor winds or unwinds the steel cord. As the cord is winded, both bands inflect and the gripper grasps. An absolute rotation sensor measures the motor revolution, which roughly corresponds to the gripper "finger" distance. This mechanical design is inspired by 600 years old chart of the robotic gripper [7].

Each gripper is attached to a carriage on a vertical rail and can be positioned between 1 cm and 22 cm above field level. The vertical rail is driven by a modified servo with IRC sensors. A switch at the upper end of the rail allows to assess, whether the gripper has reached the maximal height.

The manipulator control board is based on two ATmega microcontrollers. These microcontrollers generate signals for both grippers and are responsible for

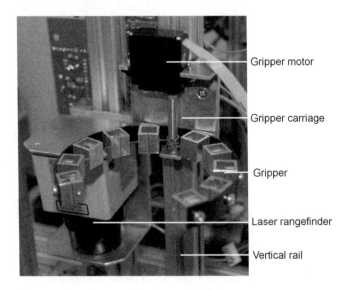

Fig. 4. Manipulator detail

their vertical positioning. Moreover, this board controls the vertical position of the main robot sensor, the laser rangefinder. The control board is interconnected with the onboard PC via a RS232 interface. Apart from printing its own status on the LCD display, it acts as a bridge between the LCD display and the onboard PC. Microcontrollers use internal interrupts and timers for better communication and event handling.

The manipulator is designed as a fairly independent function module capable of its own actions. During the item pickup, the rangefinder is positioned low to detect column elements. When a column position is reported to be in a good position, the gripper attempts to grasp it and reports if the operation has been successful. The same applies for vertical positioning - the central computer announces the height a gripper should be raised to and the control board reports completion.

5   Sensors

The task of the sensing system is to estimate positions of the construction elements and both robots relative to the playfield and therefore to prevent our and opponent damage. To achieve these tasks, we fuse data from odometry, bumpers, laser and infrared rangefinders.

5.1   Self Localization

Most of the time, the self localization is based on odometric data while bumpers and the laser rangefinder are utilized scarcely. The odometric system is based on a pair of towed wheels located close to traction wheels. The towed wheels are connected by a belt pulley to larger wheels with IRC sensors. Pulses of these sensors are counted by an independent odometry board, which converts them to Cartesian coordinates and heading. The onboard PC can obtain these data from the odometric board on request.

Although the passive odometry does not tend to lose track of the robot position even in case slipping or collisions, it is still subject to drift. Furthermore, knowing opponent robot position is useful - and this cannot be measured by the odometry.

Therefore we use an independent localization sensor, which is based on the URG-04LX laser rangefinder mounted on a vertical positioning device. Its scanning plane is parallel with the field plane and its height above the field level is adjustable in range of 2-20 cm. The sensor range is limited to 4 m, its resolution is up to 1 mm, precision is approximately 1 cm. Although its field of view is supposed to be approximately 270 degrees, it is not fully utilized because a part of the scan is obstructed by robot components. One complete scan takes approximately 0.1 s and provides 720 distance measurements.

The algorithm fusing odometry and laser rangefinder data works as follows: The laser rangefinder is lowered to 4 cm above the field. Rangefinder data are requested and the odometry board is asked for Cartesian coordinates.

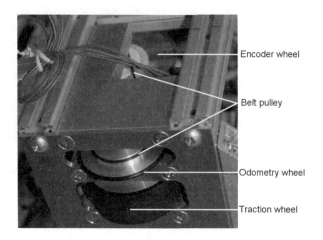

Fig. 5. Odometry wheel detail

The position determination is established on basic data pattern recognition (corresponding to walls and central building zone). After the scan is taken, significant patterns are identified in the laser data and their positions relative to robot are estimated. In the next step, correspondences between those pattern and field objects are established and several hypothesis of robot position are computed. Since localization results are ambiguous, a result closest to odometry readings is chosen. This result is reported to motion and odometry control board to correct their position information.

Two frontal mechanical bumpers serve not only to indicate successful docking to building areas, but also as a last resort collision avoidance. They are directly wired to the motion control board, which inhibits robot forward movement in case the bumpers are pressed. When our robot moves backwards, distance measurement of its rear infrared sensors are taken into account. Collisions can be also detected by comparing odometric information from passive and traction wheels.

5.2   Object Recognition

Five types of objects are recognized in the scan: columns, lintels, central building zone, walls and opponent robot. Each type of object is recognized by different algorithm. Circular object recognition (i.e. columns and central area) is based on Hough transform [8]: First of all, a 2D grid representing robot surrounding is prepared (cell size is 1 cm, grid size is 100^2 cells), i.e. all its cell values are set to zero. Then, each value of the rangefinder scan is converted to a point in the field coordinate frame. Points which lie outside the field or too near to field edges are erased and remaining points are transformed to the coordinate frame of the prepared grid. For each of these points a set of grid cells with distance of the searched object radius is found and values of these cells are increased. After all points are processed, the grid is searched for local maxima which correspond to the centers of detected circular objects. Both column elements and central zone have known radius and therefore can be be found in $O(nk)$, where n is the size of the rangefinder data vector and k corresponds to grid resolution. As an alternative, circular elements are searched by finding local minima in the rangefinder scan, pre-filtered by sliding average. The neighborhood of each minimum is then examined to decide, whether the minimum represents a column or acropolis. Though less precise than Hough transform, this alternative algorithm should be faster and more robust to noise. Currently, both of these algorithms are used in robot sensing module and their robustness is being evaluated.

Walls are recognized by Radon transform [9] in a similar fashion to circular object detection.

The lintel recognition is performed as follows: Laser rangefinder measurements closer to each other than the lintel length are grouped. Mean and covariance matrix of each such group is computed and a ratio of covariance matrix eigenvalues is calculated. A group with a ratio higher than a predefined threshold is supposed to represent a straight segment and therefore correspond to a lintel. After determination of each group mean and covariance eigenvectors, computation of position and orientation of the lintel is trivial.

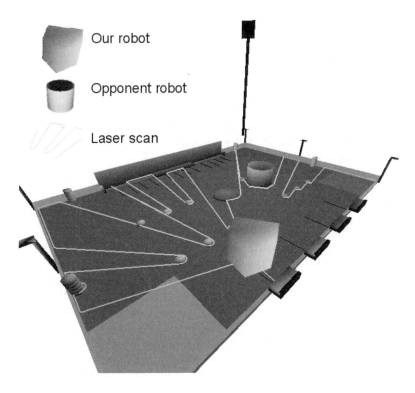

Fig. 6. Scanner data interpretation

To recognize the opponent robot position, all measurements laying inside the field, which do not represent construction elements or central area are segmented. The centroid of the largest segment is considered to represent the opponent robot.

5.3 Building Area Surveillance

The rangefinder scanning height can be adjusted to scan for objects on the building areas. This is useful in cases, when we want to place new construction elements on a previously build temple. The scanner is elevated while the robot approached the building zone. As the robot is getting close to the building area, construction elements are recognized in the scanned data and their positions relative to the robot are computed. The robot can therefore precisely position construction elements on our own or opponent building. Moreover, the scanned data allow to build a 3D model of temples at the construction site.

6   Collision Avoidance

The purpose of the collision avoidance module is to adapt the robot movement to the opponent robot position and speed. We presume, that most robots will meet only at the construction site and therefore the major part of collision avoidance can be implemented in the planning module. When deciding at which area to build, the opponent position and speed is assessed and if it is moving towards a central building area, our robot will try to build at the side construction zone and vice versa. During any movement, the opponent position is evaluated and if it is close to our robot, our movement is slowed down and eventually stopped. If our robot does not move due to the opponent presence for 3 seconds, the obstacle avoidance module is activated. The avoidance is based on a simple set of rules taking into account both opponent and our position.

7   Planning

The planning is a process of creating a sequence of actions, which we have to apply to a system in order to transfer it from its start state to the goal states. Planning can be realized as a search through the space of all possible action sequences, while looking for the optimal plan according to the given optimality criterion. In our case, the plan is a series of robot actions with the optimality criterion given by game scoring rules. These are based on number, height and shape of the temple models build during a ninety-second time interval.

We do not utilize automatic methods for finding an optimal plan, instead of this we try to propose suitable strategies leading to the maximum score.

Our team has developed a fixed plan of actions, which should deal with likely situations (e.g. collision with other robot, occupied building area). We decided to utilize Petri net [5] model, because it provides a clear and effective representation

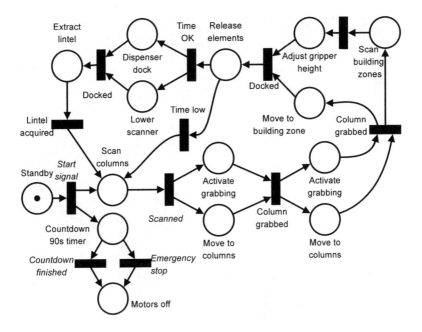

Fig. 7. Intended plan represented by a Petri Net

of the proposed plan. The Petri net is a bipartite graph consisting of places, transitions and oriented arcs. Every place in Petri net has a positive integer number of tokens. The projection from a set of places to an integer set is called "marking" and represents a state of a modeled discrete system. A transition has input and output places. For input place, there exists an arc running from it to the transition and vice versa. If there are enough tokens in every input place of a transition, this one can be "fired". During firing, tokens from input places are removed and appear in output places. The number of removed and added tokens is determined by the arc weight. The major advantage of Petri nets is the ability to model parallelism and synchronization.

In our approach, a place in Petri net represents particular functionality of robot subsystems mentioned in previous sections. These functions can be thought as simple robot behaviors [10]. If a token is present in a place, corresponding behavior is activated. There can be more than one behavior active at the same time, but within one subsystem, only one behavior is in effect. Because of Petri net formalism, we can easily examine important properties (deadlock, reversibility, liveness) of our plan. We also measure the time needed for each action and use timed Petri nets extension to estimate plan duration. This can give not only quantitative evaluation of our robots ability to win, but also shows functions, which have a major impact to the plan length. During robot tuning in later stages of development, we can optimize these behaviors to obtain greater speed.

Fig. 8. The competing robot

The intended plan represented by Petri Net is shown on figure 7. One should consider, that some behaviors are always active (e.g. reactive collision avoidance) and therefore need not be included in the Petri Net representation.

8 Conclusion

Since most of the electronics and software was reused from previous years, building the robot did not take as much time. At the finalization time of this paper, the robot is completed and its components have been tested. All of the mechanical components of the robot have been completed and control boards have been programmed.

The robot has won the Czech national round employing a simple, but reliable strategy shown on figure 7. With this strategy, the robot first attempted to construct a temple consisting of two column elements and one lintel. After finishing this "minimal" temple, the robot searched for other two columns and placed these on the previously build temple.

Before the international round, we will increase the reliability of both grippers by increasing the friction of their inner side. Vertical positioning speed of the grippers and rangefinder will be increased. Although the passive odometry has greatly increased localization precision, it requires further improvements, which should increase its robustness. The mechanics of the traction motors also requires several improvements to increase its precision and reliability. We would like to implement a simple diagnostics subsystem, which would detect failures of localization, manipulator and drive subsystems (e.g. communication mismatches, loose traction wheel, wrong localization results, nonsensical sensory data patterns). In case a failure will be detected, the corresponding module will be either reinitialized or shut down completely. The planning module will be modified to take more attention to the remaining match time. An alternative localization algorithm of circular objects 5.2 seems to give better results, but further investigation is needed.

Although the aim of this project was to create a robot capable of performing a task given by match rules of EUROBOT 09 competition, its main goal is educational. It should teach students to work as a team, evolve their knowledge in robotics and apply theoretical principles they have previously learned. Most of the current team members have participated in previous years and one can clearly observe that their labor effectivity has increased significantly. The junior team members have acquired a student research grant and therefore can manage the project almost on their own. However, the team lacks new student members and therefore should put more effort in its presentation.

A secondary goal of this project, i.e. range-based object recognition with constrained hardware has been implemented and tested in real-world conditions.

Acknowledgements

Construction of this robot would not be possible without support of Dr. Libor Přeučil, head of IMR, FEE CTU in Prague. We thank EUROBOT organizers, volunteers and team members for their efforts in contest organization. The presented work has been supported by the Ministry of Education of the Czech Republic under program "National research program II" by the projects 2C06005, MSM 6840770038 and by the Czech Technical University student grant 1/1/2009.

References

1. Eurobot Association (2009), http://www.eurobot.org/
2. Federation of International Robosoccer Association: FIRA web pages (2009), http://www.fira.net/
3. Eurobot Association: Competition rules for year 2009: Temples of atlantis (2009), http://www.eurobot.org/eng/rules.php
4. Hokuyo Automatic co.ltd.: Laser rangefinder urg04lx (2009), http://www.hokuyo-aut.jp/02sensor/07scanner/urg_04lx.html
5. Petri, C.A.: Kommunikation mit Automaten. PhD thesis, Institut für instrumentelle Mathematik, Bonn (1962)

6. Barton, M.: Controller development and implementation for path planning and following in an autonomous urban vehicle. Master's thesis. University of Sydney, Sydney (2001)
7. Rosheim, M.E.: Leonardo's Lost Robots. Springer, Heidelberg (2006)
8. Duda, R.O., Hart, P.E.: Use of the hough transformation to detect lines and curves in pictures. ACM Commun. 15, 11–15 (1972)
9. Deans, S.R.: The Radon Transform and Some of Its Applications. Krieger Publishing Company (1993)
10. Arkin, R.C.: An Behavior-based Robotics. MIT Press, Cambridge (1998)

Real-Time Door Detection Based on AdaBoost Learning Algorithm

Jens Hensler, Michael Blaich, and Oliver Bittel

University of Applied Sciences Konstanz, Germany
Laboratory for Mobile Robots
Brauneggerstr. 55 D-78462 Konstanz
{jhensler,mblaich,bittel}@htwg-konstanz.de
http://www.robotik.in.htwg-konstanz.de/

Abstract. Doors are important landmarks for robot self-localization and navigation in indoor environments. Existing algorithms of door detection are often limited for restricted environments. They do not consider the diversity and variety of doors. In this paper we present a camera- and laser-based approach, which allows finding more than 72% doors with a false- positive rate of 0.008 in static testdata. By using different door perspectives form a moving robot, we detect more than 90% of the doors with a very low false detection rate.

1 Introduction

For an indoor environment, doors are significant navigation points. They represent us the entrance and exit points of rooms. Doors provide stable and semantically meaningful structures for the robot localization. Therefore, robust real-time door detection can be used for many important robot applications. Among these applications are courier, observation or tour guide robots.

Fig. 1. Shows the mobile robot Pioneer2DX. It is equipped with a camera, a 2D laser range finder and an odometer.

In the past, the problem of door detection has been studied several times. The approaches differ in the implemented sensor system and the diversity of environments and doors, respectively. For example, in [6] and [3] only visual information was used. Others, like [1] apply a additional 2D laser range finder and receive thereby better results.

A. Gottscheber, D. Obdržálek, and C. Schmidt (Eds.): EUROBOT 2009, CCIS 82, pp. 61–73, 2010.

Fig. 2. Illustrates typical door images, taken by the Pioneer2DX camera. The top of the doors are occluded and also the diversity of doors are recognizable: The doors have different poses, colors, lighting situations, as well as different features e.g. door gap or texture on the bottom.

From this approaches, we found, that there are two major difficulties in autonomous door detection. Firstly, it is often impossible to get the whole door in a single the camera image. In our scenario, the robot camera is close to the ground, such that, the top of the door is often not captured by the robots camera (see figure 1 and 2).

The second difficulty is characterized by the multiplicity and diversity of doors (even for the same door types) in various environments. Like figure 2 depicts, doors can have different poses, lighting situations, reflections, as well as completely different features. The main features, which characterize a door are illustrated in figure 3. A door can be recognized e.g. by its color and texture with respect to the color and texture of the surrounding wall, respectively. Also door gaps or door knobs are indicators of a door. Even though these features are not always present or cannot always be detected in a single camera image, a robust algorithm should detect the door by using the remaining features.

In a recent work [3], this two issues of door detection were solved by extracting several door features out of the robots camera images and by applying the AdaBoost algorithm [5]. The algorithm combines all the weak features of a door candidate to receive a strong door classifier, which allows to decide if a door is found.

For our situation this approach was not sensitive enough. We did not reached the same high detection rate in our complex university environment with a similar system. Therefore, we added a laser based distance sensor. Further weak classifiers were used to improve the detection results. In the experimental result section we demonstrate the performance of our system on a large database of images from different environments and situations.

2   The AdaBoost Algorithm

The concept behind all boosting algorithms is to use multiple weak classifiers instead of one strong classifier and solve the decision problem by combining the results of the weak classifiers. It is often easier to find several rules of thump for a problem instead of one complex rule to solve it. The AdaBoost algorithm is

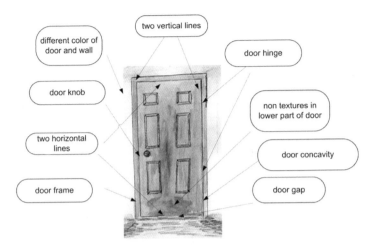

Fig. 3. Door is characterized by several features

using a training dataset to build up the strong classifier. For this purpose, the algorithm requires, that each weak classifier reaches in the training process at least 50% success and the errors of the classifiers are independent. If this is given, the algorithm is able to improve the error rate by calculating optimal weight for each weak classifier. If $y = h_n(x)$ is the output of a nth weak classifier to the input x and α_n the weight for the nth classifier created in the training process. The strong classifier is given by:

$$H(x) = sign\left(\sum_{n=1}^{N} \alpha_n h_n(x)\right) \tag{1}$$

3   Detection of Door Features

As indicated in figure 4, we use the robots camera image and the laser based distance signal for the door detection. Out of the camera image we extract vertical lines to find door candidates in the images. Hereby, we assume that each door has a vertical line on the right and left side. As a consequence, a door is not detected, if the door posts are not visible. In the next step we check each candidate for the seven door features, which represent the weak classifiers: a door can have a certain width, the color of the door can be different from the color of the wall, a door may have a door frame or door knob, the door can stick out of the wall or not. The result is not effected by a single classifier as long the weighted majority of the other classifier overcomes this loss.

3.1   Vertical Line Pairs

As mentioned before we use the vertical line pairs generated by the door frame as door candidates for the AdaBoost algorithm. We use the Contour Following

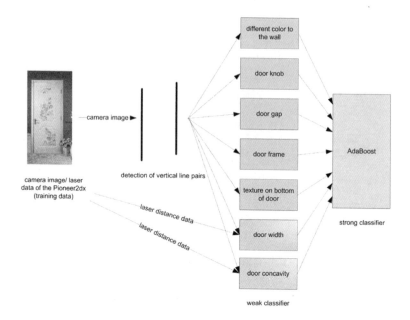

Fig. 4. This Image illustrates our approach to find doors. Several weak classifiers are combined using the AdaBoost algorithm in the training process. The door width and door concavity classifiers are using the laser distance data from the robot in addition to the camera image.

Algorithm [7] to find vertical lines. Compared to other transformations, this method has the advantage, that we obtain the start and end points of the lines.

Figure 5 shows the typical result of our line detection algorithm. Broken contours caused by door hinges or knobs were overcome by merging vertical lines separated by a small gap. Reflection of the door on the ground causes unexpected long lines, which lead to failures of some weak classifier. Those long lines were eliminated by cutting lines on the ground and optimizing the Canny Edge Detection algorithm.

Not every combination of vertical line pairs in an image correspond to door candidates. The number of candidates was drastically reduced by the following rules:

- The vertical lines need to have a minimal length.
- The horizontal distance between the two vertical lines has to be between a maximal and minimal value.
- The end points of the vertical lines have a minimal vertical shift.
- If there are lines close together, which all may represent a door candidate according to the earlier rules, only the inner lines are used. The other lines are indicators for a door frame.

Fig. 5. Vertical line pairs after the Contour Following Algorithm. Two vertical line pairs according to same door candidate. Some combinations can be excluded. Line pairs corresponding to a door candidate are labeled with the same green number in the image.

3.2   Door Gap

To avoid friction between the door and the ground, doors are constructed with a small gap. As a result, this gap appears brighter or darker than the surrounding area, depending on the illumination of the room behind the door. With a look at vertical profile we determine the maximum or a minimum in the vertical profile (figure 6). For each pixel along the bottom door edge, we compute the minimum and maximum intensities in the vertical profiles. If the mean value of all minimum and maximum values is above or below a certain threshold the classifier found a door.

3.3   Color of the Wall

In many environments doors have a different color than the surrounding walls. Figure 7 illustrates that the 2D hue- saturation histogram of the door is totally different to the histogram of the wall. To measure this feature we use the *histogram intersection* [8] method. Given is a pair of histograms (door D and wall W), each contains n bins. The intersection, bins with the same color, can be calculated:

$$\sum_{j=1}^{n} min\left(D_j, W_j\right) \qquad (2)$$

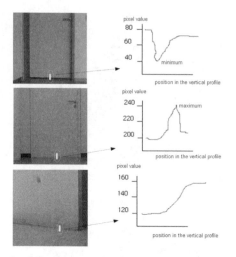

Fig. 6. This figure shows intensities in the vertical profile of a door. In the first image the light in the room behind the door is off. As a result, we receive a minimum in the vertical profile. In contrast, the light in the room behind the door in the center image is on. In consequence, the vertical profile has a maximum in the vertical profile. If there is no door gap, which is shown in the last picture, the vertical profile does not have any minimum or maximum.

For a value between 0 and 1 we normalized this formula:

$$H(D, W) = \frac{\sum_{j=1}^{n} min(D_j, W_j)}{\sum_{j=1}^{n} D_j} \qquad (3)$$

with

$$H(D, W) < T_{wall} \qquad (4)$$

and T_{wall} is the threshold which decides if a door is found or not. One difficulty here is to choose the right area for the wall and the door. Therefore, we use the distance between the two vertical lines. Additionally, we have to consider that the door frame is not included in this areas.

3.4 Texture

The bottom area of a door is usually not textured. In contrast, the wall and other objects are textured in this area. Textures are going to be visible in the Canny Edge [2] detection picture (figure 8). To compute this attribute of a door, we sum up the magnitude values $\nabla I(p)$ in the door bottom area:

$$\frac{1}{|A|} \sum_{p \in A} |\nabla I(p)| < T_{texture}, \qquad (5)$$

where p corresponds to a pixel in the area. The sum is divided by the number of the pixels $|A|$. The decision is obtained by comparing this value to a threshold

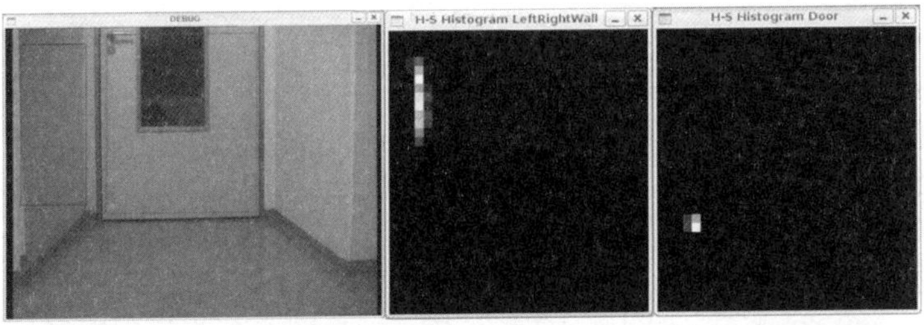

Fig. 7. Left: Areas which are considered for the ColorWallClassifier. Center: 2D hue-saturation histogram of the wall area. Right: 2D hue-saturation histogram of the door area.

Fig. 8. Shows an image after Canny Edge detection. Green area: No texture in the bottom area of the door. Yellow area: Others objects have often textures in this area.

$T_{texture}$. It is important for this classifier to consider that the area has to grow, if the distance between the two vertical lines gets larger.

3.5 Door Knob

For the door knob classifier we use again the line image calculated for the detection of the vertical line pairs. However, for this classifier the vertical lines are not important, but the almost horizontal lines, which are result from the door knob. The distance between the vertical line pairs is used to find the door knob area. In these two areas (left and right side of a door) the classifier returns a "true", if:

$$\sum (line_{almost\_horizontal}) \geq T_{doorknob} \qquad (6)$$

where:

$$T_{doorknob} = 2 \qquad (7)$$

Image 9 shows typical detected door knobs using this weak classifier.

3.6 Door Frame

The door frame is required to install a door inside a wall. The frame can be calculated during the determination of the vertical line pairs. A door frame in

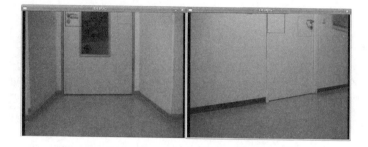

Fig. 9. Two example images, which show the door knob areas. Door knobs are detected if inside of these areas are at least two almost horizontal lines.

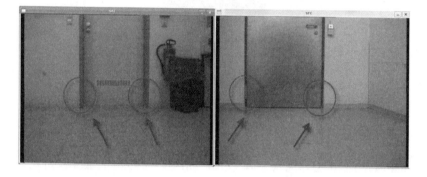

Fig. 10. The red arrows point to areas where duplicated vertical line pairs are found. The door frame classifier will be positive for these door candidates.

an image is characterized by duplicated vertical lines. If there is an additionally vertical line on both sides of the door, the door frame classifier is true. Figure 10 presents the situation.

3.7 Door Width

There is a DIN standard (DIN18101) for the door width. Unfortunately the door width can strongly vary. Even here it is not easy to find a strong classifier for a door. But for a weak classifier we bordered the width in values between 0.735m (threshold for minimal width) and 1.110m (threshold for maximal width). To calculate the distance between the two vertical lines we use the laser distance data provided from the robot.

3.8 Door Concavity

In many environments doors are receded into the wall, creating a concave shape for the doorway (illustrated by figure 11). This shape can be obtained using the robot laser distance data. For that the slope between each measured laser distance in the door candidate area is calculated. Figure 12 shows the slope

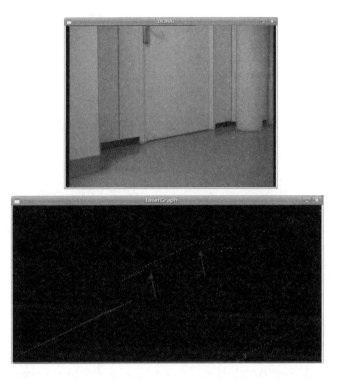

Fig. 11. The red arrows in the laser profile point to a door. The images show that the door is not running with the wall. It is receding or sticking out from the door.

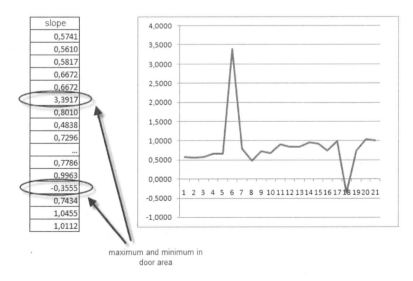

Fig. 12. Slopes between the measuring points from figure 11. We found turning points in the area of the door frame.

between the different points for the sample laser profile. There exists a maximum and minimum at the position of the door candidate. The Concavity or U-Shape weak classifier can be described by the following rules:

- If we calculate the slope between each measured laser distance point, without considering the door candidate, the standard deviation is almost zero ($T_{standard\_deviation}$).
- If we look at the slope at the door frame area we will find values which strongly vary from the calculated mean value.

4   Experimental Results

To test the performance of the system, a database of 210 test dates were taken with the Pioneer2DX. One test date consists of one camera image, one laser profile as well as the position of the camera at a certain time. We considered pictures of doors and non-doors. From the 210 test dates we took 140 for the training process of the AdaBoost. The residual of the 210 test dates were taken to test our system. In these 70 test dates our door candidates finder algorithm found overall 550 door candidates, where 61 candidates correspond to a real door. The result is shown in a ROC chart [1] (figure 13). As we can see in the chart, the AdaBoost classifier does not reach the best detection rate. In our test the true-positive rate for the AdaBoost classifier reaches a value of 0.72 and a false-positive rate of 0.0082. However, for an object detection system, where a small false-positive rate is more important than a high true-positive rate, the result of the AdaBoost classifier is the best. With a large probability the door will be detected anyway from the robot form another robot position. But a false detection would cause an unwanted detection of a door. We get the same result, if we take a look at the RPC methods [2] (table 1). The best value of the F-score (combination from precision and recall) is obtained by the AdaBoost classifier.

Table 1. Result of the RPC methods. The F-score can be interpreted as a weighted average of the precision and recall, where a F-score reaches its best value at 1 and worst score at 0.

	Recall	Fallout	Precision	F-score
DoorWidthClassifier	0.9016	0.2413	0.31791	0.470
DoorConcavityClassifier	0.5737	0.1554	0.3153	0.4069
TextureBottomClassifier	0.7541	0.0920	0.5054	0.6052
ColorWallClassifier	0.0655	0.0348	0.1904	0.0975
DoorGapClassifier	0.5901	0.3149	0.1894	0.2868
DoorKnobClassifier	0.8988	0.1022	0.5145	0.6463
DoorFrameClassifier	0.4426	0.2965	0.1569	0.2317
AdaBoostClassifier	0.7213	0.0082	0.9166	0.80733

---

[1] Receiver Operating Characteristic.

[2] Recall Precision Characteristic.

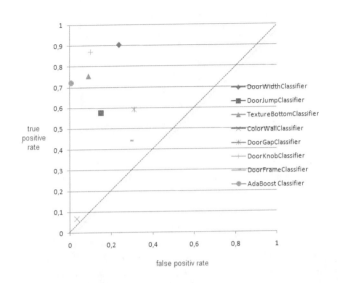

Fig. 13. ROC diagram of all classifiers. The best value is reached at coordinate (0,1). The AdaBoost classifier does not reach the best detection recall ration. However, for an object detection system, where a false detection could have undesired effects, the AdaBoost classifier is the best classifier.

Fig. 14. Typical by the AdaBoost classifier detected doors. The pictures demonstrate that our approach is robust against different robot positions, reflection situation and as well as different door features.

Fig. 15. Picture illustrates a sample false-positive error of the AdaBoost classifier. In the sample a wall, which looks similar to a door, is detected as door.

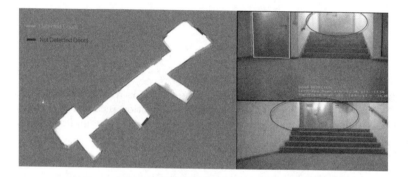

Fig. 16. The Image depicts the result of the robot test run in the basement environment. Detected doors are marked in the map. The doors that have not been detected by the algorithm (red ellipses in the second image) are in position which could not be measured by the laser.

Typical detected doors are illustrated in figure 14. As can be seen, the algorithm is capable of detecting doors under different lighting situations and different viewpoints of the robot. Also the absence of one or more door features does not cause a non detection of the door.

False detections happen by candidates, which have a very similar appearance like a door e.g. in figure 15 a wall is detected as a door. In this example the DoorWithClassifier, DoorJumpClassifier, TextureBottomClassifier and the DoorKnobClassifier are positive.

In a next step we tested the system as a Player [4] driver on our Pioneer2DX robot. Hereby, we used two different environments. We also received a good result. In the first environment (basement of the university) all doors were detected, except two, which could not be measured by the laser system of the robot. The result of the first environment is shown in the figure 16. In the second environment (office environment) each door, beside glass doors, were detected. The problem here is, that we received wrong laser distances because the laser is going through the glass. Also the reflection on the ground of these doors is higher than on other doors.

5 Conclusion and Future Work

In this paper we presented an approach for a laser and camera-based door detection system. By using the AdaBoost algorithm we created a system with a detection rate with more than 72% and a very low error rate of 0.008. It is a combination of several weak classifiers e.g the color of the wall, door knob or door gap. In the result section we used the ROC and RPC methods to demonstrate that none of the other weak classifier can replace the strong classifier created by the AdaBoost algorithm. The system has the ability to find doors in realtime. With an Intel Core Duo 2.4GHz processor we reached a performance of 12fps. There are multiple possibilities to improve the system. Firstly, the training set can be enlarged. More test data would improve the alpha values for each weak classifiers. If we use the system in a new environment, it will provide a better result, if we add test data of this environment. Secondly, the weak classifiers can be modified and new weak classifiers can be added. E.g the ColorWallClassifier can be improved, if the system automatically learns the wall color of the environment. New classifiers could use the door hinges or the light switch on the door side. For future work it would be interesting to integrate this system in an autonomous map building system. That means that the robot has the ability to create a map of an unknown environment and mark doors in it. Secondly, the detection of a door plate would also be interesting to navigate the robot better through unknown environments. Furthermore, the detection of open doors with new classifiers would be interesting.

References

1. Anguelov, D., Koller, D., Parker, E., Thrun, S.: Detecting and modeling doors with mobile robots. In: Proceedings of the IEEE International Conference on Robotics and Automation, ICRA (2004)
2. Canny, J.: A computational approach to edge detection. IEEE Trans. Pattern Anal. Mach. Intell. 8(6), 679–698 (1986)
3. Chen, Z., Birchfield, S.T.: Visual detection of lintel-occluded doors from a singel image. IEEE Computer Society Workshop on Visual Localization for Mobile Platforms 1(1), 1–8 (2008)
4. Collett, T.H.J., MacDonald, B.A., Gerkey, B.: Player 2.0: Toward a practical robot programming framework. In: Australasian Conference on Robotics and Automation, Sydney (2005)
5. Freund, Y., Schapire, R.E.: A short introduction to boosting. J. Japan. Soc. for Artif. Intel. 14(5), 771–780 (1999)
6. Murillo, A.C., Košecká, J., Guerrero, J.J., Sagüés, C.: Visual door detection integrating appearance and shape cues. Robot. Auton. Syst. 5
7. Neira, J., Trados, J.D.: Computer vision. In: Universidad de Zaragoza, Zaragoza, Spain (2008)
8. Swain, M., Ballard, D.H.: Color indexing. International Journal of Computer Vision 1(1), 11–32 (1991)

DC Motor Drive for Small Autonomous Robots with Educational and Research Purpose

Damir Krklješ, Kalman Babković, László Nagy,
Branislav Borovac, and Milan Nikolić

Faculty of Technical Sciences, Trg Dositeja Obradovića 6, 21000 Novi Sad, Serbia
{krkljes,bkalman,lnadj,borovac,mnikolic}@uns.ns.ac.yu

Abstract. Many student robot competitions have been established during the last decade. One of them, and the most popular in Europe, is the European competition EUROBOT. The basic aim of this competition is to promote the robotics among young people, mostly students and high school pupils. The additional outcome of the competition is the development of faculty curriculums that are based on this competition. Such curriculum has been developed at the Faculty of Technical Science in Novi Sad. The curriculum duration is two semesters. During the first semester the theoretical basis is presented to the students. During the second semester the students, divided into teams of three to five students, develop the robots which will take part in the incoming EUROBOT competition. Since the time for the robot development is short, the basic electronic kit is provided for the students. The basic parts of the kit are two DC motor drives dedicated to the robot locomotion. The drives will also be used in the research concerning the multi-segment robot foot. This paper presents the DC motor drive and its features. The experimental results concerning speed and position regulations and also the current limiting is presented too.

Keywords: DC motor drive, autonomous robots.

1 Introduction

One of the most attractive multidisciplinary fields of investigation, research and employment is certainly robotics. Robots have entered almost all fields of human living. This ongoing process promises even more and more robotisation of the word, opening many new subfields of robotics. To keep this process going we must investigate how to attract and educate young people. The easiest way to attract young people is through interesting robot competitions. Many universities, faculties, schools, organizations have recognized this, and have established popular robot competitions, targeting young people. A brief survey of these competitions, covering almost all, can be found in [1]. One, that is unjustifiably omitted is the European competition EUROBOT. Since the source for this is the Wikipedia, a user developed and edited encyclopedia, this will soon be corrected. There is an unbreakable link, almost a symbiosis, between the education in the field of the robotics and robot competitions. In order to advertise robotics and attract young people robot competitions are

A. Gottscheber, D. Obdržálek, and C. Schmidt (Eds.): EUROBOT 2009, CCIS 82, pp. 74–87, 2010.

established. Backward influence of the competitions is in their impact on the curriculum, covering the robotics topics, focusing it to well defined, not abstract, problems that must be addressed and solved. There is nothing more useful to the students and pupils then to be evolved in practical, multidisciplinary team projects. This fact cannot be emphasized enough. The outcomes for the students being engaged in such projects can be found in [2]. The links and relations between the education in the field of robotics and robot competitions are covered in [3,4]. A brief introduction to RoboCup, including the basic rules, the competition environment, and typical research work over RoboCup can be found in [5].

Following the trend to develop a curriculum in robotics that relies on robot competitions, a new curriculum has been developed at the Faculty of Technical Sciences in Novi Sad, Serbia. The curriculum is devoted, but not restricted, to the Mechatronic branch of the faculty. The curriculum takes part at the fourth year of studies and lasts for two semesters. In the first (winter) semester the students gather theoretical knowledge in the fields of electronics, mechanics, sensors, actuators and microcontroller programming, focusing to the parts of these fields relevant to robot design and construction. The second (summer) semester is devoted completely to the robot construction. The students form teams of three to five students. Each team has a tutor who is responsible for providing guidance and consultation. The final outcome of the second semester, for each team, is a small autonomous robot that will respond to the current year task and competition rules. The competition that this curriculum currently relies on is the European robots competition called EURUBOT (http://www.eurobot.org/). In this competition autonomous robots compete to win in achieving a defined task. Each year the task is different, but usually the dimensions of the playground remain the same. The tasks are usually some kind of picking, sorting, collecting and deploying of some known objects. The competition usually takes place in the second half of May. National competitions (if there are more than three national teams) must be held before, and the three best ranked teams are eligible to participate in EUROBOT competition.

Since the time for the robot development is short, the basic electronic kit is provided for the students. Additional electronics, like sensor circuits, are left to them to develop. The basic parts of the kit are two DC motor drives dedicated to the robot locomotion. The drives described in this paper are improvements of the previous drives [6].

The purpose of the drives is not limited to the small robot locomotion. Our intention is to use the drivers in laboratory exercises that concerns DC motors and drives in the "Application of Sensors and Actuators" curriculum. The drives will also be used in research in the field of biped robotics. The research will include the actuated multi-segment robot foot such as the one mentioned in [7].

2 Educational Aspects

As mentioned before, our main intention is to provide a good starting point in robot development and also to shorten the development time. Although it might seem that the students will not benefit in knowledge, as they would, if the development was left to them that, it is not the case. First of all, during the first semester, the basic principles of different kinds of motor drives are presented. Their major characteristics concerning the output stage topology, insulation, switching mode drive circuits, circuit component choices, current limiting and prevention of shoot-through currents are explored. The demands on microcontroller for the applications of motor control are introduced. Then the DC motors as objects of regulation are explored. Basics of digital regulation are then introduced. After that, previous solutions of motor drives are analyzed, highlighting their good and bad properties. With that knowledge they are capable of understanding the motor drive that they are going to use. The DC motor drive is then introduced and explained in details. Practical exercises and demonstrations then further improve their knowledge and capabilities to use the drive correctly. A handbook for the drive was written. This handbook provides all the details about driver construction, external connections, working principles and communication parameters. During the second semester, in which the students apply the drive, guidance and consultation is provided by the team mentors. The students are also encouraged to interchange their experience among teams as well as within the teams.

We believe that this curriculum gives good theoretical and practical knowledge and beside that the students are introduced to team work. Practically the students are put in to situations in which they will find themselves during their employment after the studies. We have recognized the importance of the team work and paid attention to that aspect. A few lectures, lectured by an expert in the field of team work, are also organized, providing the students with the basic knowledge in that field.

3 Driver Design

The starting point in any design is to define all design requirements based on its purpose and other demands. In our case the drivers are dedicated to the autonomous robots that will be used by the students. From that input we can start to define the properties of the driver.

3.1 General Considerations

Before thinking about a particular solution for the drive some general properties must be posed. The consideration is based on the type of user and particularities of a task to be performed. Our users are students. We must suppose that they are not practiced enough. Therefore the design must be strait and clear to them and also presented in a comprehensive way. Simple manipulation and implementation of the driver must be achieved. That means simple and clear hardware connections and software interface to the driver and simple configuration of the driver. Each robot has different properties (dimensions, weight, transmission...), therefore the properties of the driver software components, especially those that affect regulation, must be adjustable. A protection from the driver accidental misuse must be implemented. Malfunctions of the driver should be easily detected, located and repaired. An objective is also to use cheaper and easily obtainable electronic components.

3.2 Power Demands

The first choice made is concerning the output power of the drive. The inputs upon which the choice will be made are: a maximum robot weight (m_{max}), a maximum robot speed (v_{max}) and a maximum acceleration (a_{max}). These inputs are determined from experience from the previous competitions. They are m_{max} =10kg, v_{max} =2m/s and a_{max}=3m/s^2. From this, the maximum output power is:

$$P_{max} = F_{max} \cdot v_{max} = m_{max} \cdot a_{max} \cdot v_{max} = 60W . \tag{1}$$

Supposing that the efficiency coefficient is 0.6 and knowing that two motors drive the robot, we end up with the needed electrical power per motor (driver) of 50W. Concerning that the drive is powered from a 12V battery, the maximum average motor current is 4.2A. Due to the fluctuations of the motor current (PWM drive) the current limit of the drive is set to 5A.

Based on these results the choice of power electronic components was made. Also the current limit was defined.

3.3 Output Stage Topology

The output stage topology can be of two kinds. The first is a linear and the second is switching. The usual boundary between these two topologies is made on the rated output power. This boundary is usually set to 100W. The linear topology is usually used up to this boundary. There are also other criterions to choose between topologies. In our case these criterions are the insulation of control electronics from the output stage and the physical dimensions of the driver. The insulation is usually achieved either through optocouplers or transformers. When using optocouplers it is more complicated to transfer an analog than a digital signal. When using transformers it is practically impossible to transfer analog signals. The linear output stage dissipates much more power than the switching. Therefore a large heat sink must be applied. This will lead to the increase in the overall driver dimensions. Based on this investigation a switch-mode output stage was chosen.

3.4 Output Stage Components

After defining the output stage topology the output stage electronic components selection has to be made. Since a switching topology is used and needs to drive the robot in both directions an H-bridge topology was chosen. Integrated H-bridge circuits available on the market are either expensive, hardly obtainable or their characteristics are not suitable for our design. Therefore the choice was to make an H-bridge composed of discrete components. One of the suitable choices is BUZ11 [9]. The driving circuit for the power mosfets was chosen based on the following criterions: that it has a shutdown function (to be used for a current limiting), that it can drive high side N-channel mosfets and that the propagation delays are well matched. We have chosen IR2110 [10]. This circuit can also be interchanged with IR2112.

3.5 Insulation between Electronic and Output Stages

Every electronic circuit should be carefully designed in order to minimize influence of noise and interference. Circuits that combine low and high power electronic parts are critical. Interference is mostly induced in the low power parts of electronics from the high power parts. This may cause an abnormal behavior of the circuit or complete malfunction. Therefore, measures of percussion must be taken. Those measures usually refer to the printed circuit board (PCB) especially the way the common node (which is the ground reference) is routed. In case when the circuit combines low and high power parts this is usually done by separating the low and high power grounds and having different power supplies for both parts of the circuit. Connecting those grounds in one point only is a one solution. This solution fails when there is more than one such circuit powered from the same power source. This is the situation that we were confronted to, because there are two motor drivers powered from the same source. All motors and output stages are powered from one battery and all the low power electronics is powered from the other battery. In this case only complete insulation between the two parts is the solution. This can be done by transformers or optocouplers. Since transformers cannot transfer dc signals and because they are more expensive, harder to obtain and space occupational, our choice is to use optocouplers. In our design there are two kinds of optocouplers referring to their operational speed. A low speed optocoupler is used to transfer a bridge off signal that shuts down all the power switches in the H-bridge. Two high speed optocouplers 6N136 [8] are used to transfer the high speed PWM signal (up to 50kHz).

3.6 Current Limiting

An essential property of any drive is its capability to limit the current through the motor. High spikes in motor current, which occur when the motor voltage is abruptly changed (like when starting the motor), could destroy either the power electronic components or the motor and its reduction gear. Continuous currents above a certain limit can permanently damage the motor. We have implemented cycle-by-cycle current limiting as the safest way. This kind of the current limiting does not increase the frequency of the PWM signal as it may happen if hysteretic current limiting is implemented. The increase in the PWM signal frequency is followed with the increase in the power dissipation of the power components. The drawback of the chosen solution is that the average current during the current limiting is lower than the limiting current, which leads to a reduced continuous motor torque. This effect is pronounced if the motor winding inductance is low or the PWM signal frequency is too low. We have raised the PWM signal frequency to an acceptable level. Still, the motors used have low inductance. To increase the inductance, an additional inductor is connected in series with the motor. This additional inductor, except for boosting the continuous torque, flattens the motor current shape. The flatter current waveform has two effects. The first is that the torque oscillations are smaller (finer regulation). The second one affects the electric and magnetic dissipation of the motor due to the lower motor current RMS. Therefore the motor operating temperature is reduced by flattening the motor current. Furthermore the conductive component of electric and electromagnetic interference is also reduced.

The drive supports cycle-by-cycle current limiting with the adjustable current limit in the range from 0.5A to 5A via a potentiometer.

3.7 Prevention of the Shoot through Currents

A shoot through (vertical) current in the H-bridge may occur if both switches in one branch of the bridge are simultaneously conductive. This current can be very harmful for the switches and other electronics. Current limiting is not very effective way to fight this problem. It might prevent destruction of the electronic components, but it cannot prevent high levels of interference inducted by the high speed current spikes in the power supply wires. The reason for the rapid changes of these currents is the small time constant due to the low inductance of the power supply wires. Therefore these currents must be prevented. The prevention is done by introducing a delay time between the signal that turns on one switch and the signal that turns off the corresponding switch in the same H-bridge branch. The delay time must correspond to the switching time of the switches. The classical solution for this usually incorporates a nonlinear resistor-capacitor-diode delay circuit and additional two buffers for each switch. This solution is component and space consuming. In our case the insulation optocouplers are used. The fact is that the rising and falling times of a standard optocoupler output signal are not equal. The time that corresponds to switching on the optocoupler output transistor is shorter. If we use that fact and additionally increase this difference, a delay time can be implemented. The simplified version of the circuit is shown in Fig. 1.

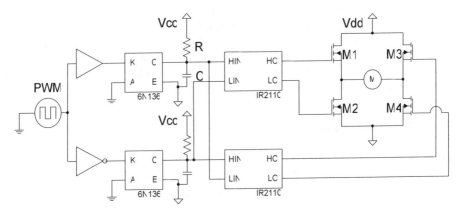

Fig. 1. Simplified schematics illustrating mosfet gate signal generation

The circuit illustrates the gate signal generation of the mosfets where the elimination of the vertical currents is achieved. Two mosfet drive integrated circuits (IR2110) are used to drive the H-bridge. Their inputs (HIN and LIN) are active high Schmitt-triggered inputs. If the enable signal for IR2110 is active (omitted in the figure), outputs H0 and L0 drive the mosfet gates in correspondence to HIN and LIN inputs. The mosfets M1 and M4 are in phase and in opposite phase to M2 and M3. The PWM signal and its inverted pair are passed through the two fast optocouplers

(6N136). Both optocouplers' output transistors are connected in common emitter configuration with collectors externally pulled-up trough resistors to the positive supply rail. Additional capacitors connected to the optocoupler collectors lengthen the rise and fall time of the output signal. They have more impact on the rising than the falling time. Since the falling edges are faster, the mosfets are turned off faster than they are turned on. This simple solution prevents the vertical currents by inserting the necessary dead time. Only two additional capacitors have been added to the original circuit.

3.8 Control Electronic

So far we have considered mostly the power part of the driver. The low power part or control electronic is usually based on some microcontroller. There are numerous criterions influencing the microcontroller choice. Our main criterions were: speed, counter and external interrupt input for encoder connection, hardware PWM generation, EEPROM memory and versatile serial communications. Beside these mentioned technical criterions very important was also the existence of obtainable free C-compiler for that microcontroller. The educational editions of commercial compilers are usually code-size restricted. The restrictions are too severe in our case. Our choice was the microcontroller of "ATmega8" type from "Atmel" manufacturer and the free C-compiler "WinAVR". We will not go further into discussion about the actual realization. We will only mention few facts about the design. The first is that the microcontroller has an external clock of the maximal 16MHz providing up to 1 MIPS/MHz instruction execution. The next one is that the standard encoder connector for the "Maxon" motor encoders is provided. The channels of the encoder are connected to the timer-counter and the external interrupt input pin in order to provide speed and position measurement for both directions of rotation. Two serial interfaces are implemented. The first one is a standard asynchronous and the second is a two wire serial (I^2C). For the convenience of programming, the ISP (In Circuit Programming) connector is provided.

4 Software Components

There are two types of software. The first one is the firmware running on the device microcontroller providing all the necessary functionality of the driver. The second one is additional and runs on PC. The PC software is provided for the convenience of driver usage.

4.1 Firmware

The firmware is written in C using free C-compiler "WinAVR". It has four modes of operation, setup through either one of the serial interfaces. The first one is "Off mode" in which no regulation is applied and all the mosfets in H-bridge are turned off. After the reset the device is in this mode. The second mode is called "Direct out mode". In this mode no regulation is applied but the output stage is turned on and the PWM signal is present. The duty cycle of the PWM is selected via the command that selects this mode. The third mode is "Speed mode" in which the velocity of the motor shaft is regulated based on the reference set by a specific command and the parameters of the *speed PID controller*. Finally, the fourth mode of operation is "Position mode" in which the position of the motor shaft is regulated. This mode uses the *position PID controller* parameters.

4.1.1 Digital Controller

The speed and position of the motor shaft is digitally controlled (regulated) using a discrete implementation of the PID regulator. The sensor for speed and position is the incremental encoder integrated during the motor assembly. The signals from the encoder are counted and the current speed and position are measured. Since the students have the opportunity to choose the motor and its associated reduction gear and also the encoder to suit the robot design best, the parameters of each PID controller must be adjustable in order to achieve the desired behavior. There are two independent PID controller structures, one for speed control and the other for position control. The parameters of each PID controller are stored in the internal EEPROM memory. These parameters are shown in Fig. 2.The figure also shows the user interface on the PC side that provides management of the PID parameters. Except for adjusting, the parameters can be saved to a file for later use. They can also be read from and written to the driver and even saved in the internal EEPROM memory of the

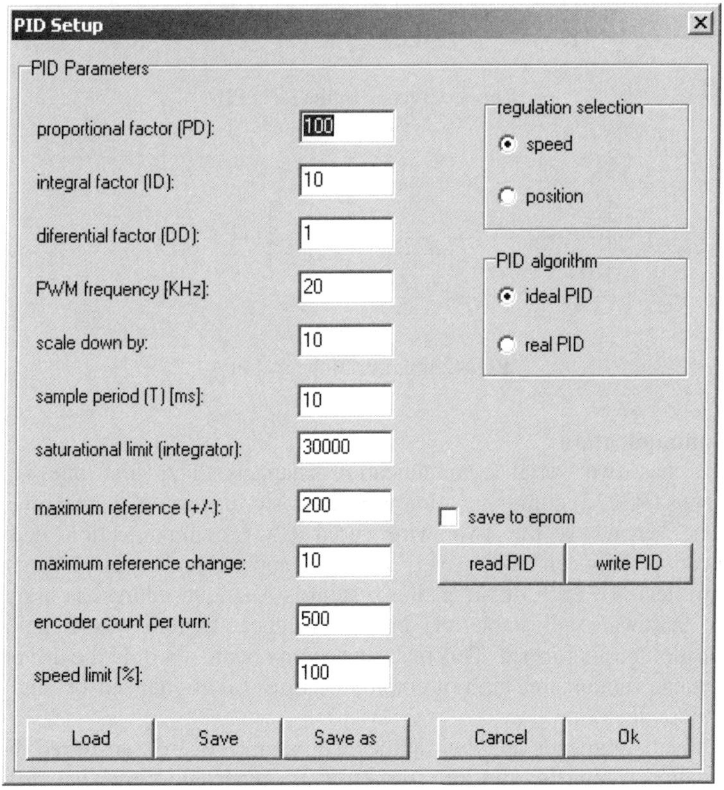

Fig. 2. Interface and parameters of the PID controller

driver microcontroller. There are two configurations for the PID controller. The first is so called ideal and the second is real PID configuration. The user can choose which one to use. The benefit of the real PID configuration is that the integrator is transferred to the output of the regulator and in that way the user should not worry about integrator saturation. The drawback of using the real PID is in greater complexity that can worsen the numerical errors. The structures of both regulator configurations are shown in Fig. 3 and Fig. 4.

The implementation of the PID regulators also supports a soft reference change. This feature is very desirable for this kind of application, when the robots are driven by the two independently controlled motors. In a situation of abrupt reference change, even if it is changed simultaneously for both motors, asymmetries of the transmission mechanics and frictions could lead to unwanted initial robot rotation, forcing it to miss the desired trajectory.

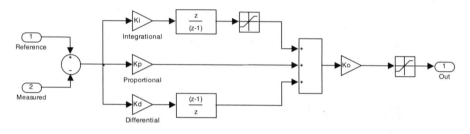

Fig. 3. Structure of the ideal PID

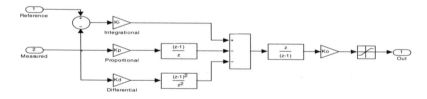

Fig. 4. Structure of the real PID

4.1.2 Communication

The driver has two serial communication channels. The first one is standard asynchronous (RS232) communication with the basic function to connect the driver to a PC. The second is the two wire serial (I^2C) communication dedicated to communication between the driver and the upper level control device (main microcontroller). To each driver in the system is a unique address is assigned. The address is hardware adjustable via binary switches. In this way a unique serial communication bus is formed. This bus can accommodate not only the drivers but any device that can support this kind of communication. Additional sensors and actuators could be those devices.

The defined commands and their associated parameters are transferred through the communication channels and the response is expected from the drivers. The communication is packet oriented incorporating start character, address, command,

command parameters, packet length, check sum and end character. In case of I²C communication the start and end characters are omitted and the address character is inherent to the communication. In case of the I²C communication the received commands are executed immediately. By using the broadcasting (addressing all drivers simultaneously) a synchronous operation of the drivers is achieved.

4.2 PC Software

A piece of software called "SoKoMo", running on a PC is developed in order to simplify the driver configuration (to setup PID parameters). The software also provides a functionality of motor response acquiring. The response is graphically presented. The response could also be saved in a numerical or picture format. The communication channel is standard asynchronous serial (RS232). The interface for the driver configuration is already shown in Fig. 2. The main software window is shown in Fig. 5. The graph shows one of previously saved responses gathered during position control. The bottom part of the window is filled with the command buttons and edits. We will explain them by an example. Suppose we want to control the position of the motor shaft according to the given response (Fig. 5).

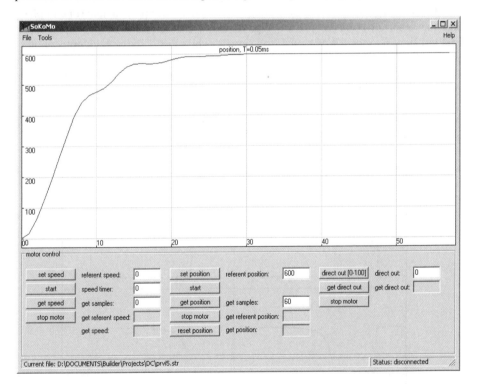

Fig. 5. Main window of "SoKoMo" software

We will assume that the PID parameters have been correctly setup. The first step is to setup the new reference by editing the *referent position* field. Then if we want to obtain the response we should fill the *get samples* field with the desired number of the response samples, in this case 60. After these steps, two mouse clicks on *set position* and *start* buttons will start the motor and end up with the response shown on the graph. During the command execution current speed and position can be obtained by clicking on the *get speed* and *get position* buttons to see how good the reference is tracked.

5 Experiments

We will illustrate the two basic properties of the driver through experiments. The first experiment will refer to the dead time generation that eliminates the shoot through currents. The second experiment will demonstrate the mosfets drive signals during the current limiting state of the driver. The example of the position response is already given in paragraph 4.2.

5.1 Dead Time Generation

The dead time generation that prevents the shoot through currents is explained in paragraph 3.8. We will refer to the Fig. 1 during the explanation. In this experiment 40kHz PWM signal is transferred through two high speed optocouplers (one in phase and the other in opposite phase). Fig. 6 shows the response of one optocoupler (bottom signal) to the PWM signal (top signal), while Fig. 7 shows both optocouplers' signal. We can see that the propagation delay is very short providing the correct transfer of the signal. Another important remark is about the falling and rising times of the optocouplers' signals. The falling time is shorter providing earlier turn-off of the currently conductive mosfets. This feature is illustrated in Fig. 8. It shows the lower mosfet drive signal (bottom signal) as a response to the optocoupler signal (top signal). There, we can see the introduced delay.

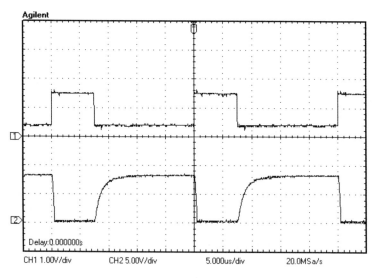

Fig. 6. Optocoupler response to input PWM signal

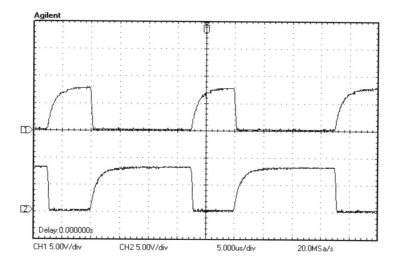

Fig. 7. Both optocouplers' response in opposite phase

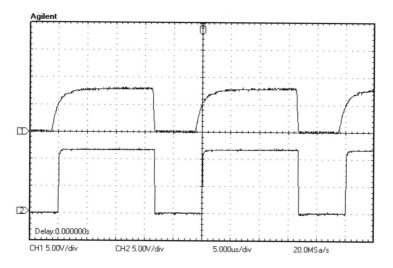

Fig. 8. Lower mosfet drive signal compared to the driving optocoupler signal

The final effect can be seen in Fig. 9. It shows two low side mosfet drive signals that are in opposite phase. We can clearly see the dead time. That is the time when both signals are low (both mosfets are turned off). In this way the shoot through currents are prevented.

5.2 Current Limiting

The current limiting is discussed in paragraph 3.7. During the experiment the 40kHz PWM signal is supplied. The driver was in the "Direct out" mode of operation. The

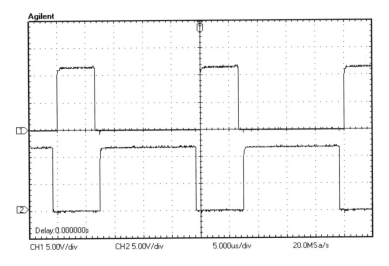

Fig. 9. Opposite low side mosfet drive signals

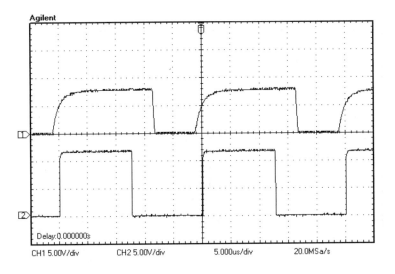

Fig. 10. Cycle-by-cycle current limiting

current limit is set to 0.5A through the potentiometer provided on the driver in order to put the driver in the current limiting state easier. Fig. 10 shows the optocoupler signal (lower) and corresponding lower mosfet drive signal in the current limiting state. Comparing Fig. 10 and Fig. 8 we can see the effects of the current limiting. The effect is shortening the mosfet drive signal. The mosfet is turned off earlier because the motor current has reached the current limit. The current limiting event turns off not only the lower mosfets but entire H-bridge. The freewheeling diodes conduct the motor current until the current disappears or another PWM cycle starts.

6 Conclusion

In this paper we have presented the DC motor drive suitable for mobile robots and also for educational and research purposes. The educational aspects were explained and a short overview of the curriculum was given. The main properties of the driver were given and they were illustrated through experiments. The further development (next driver generation) will depend on feedback provided from the students who have used the driver. The first generation of students reported only minor remarks.

Acknowledgment

This paper is one of the results of the research project: 114-451-00759/2008, financed by the Provincial Secretariat for Science and Technological Development, Autonomous Province of Vojvodina, Republic of Serbia.

References

1. Wikipedia, http://en.wikipedia.org/wiki/Robot_competition
2. Pack, D.J., Avanzato, R., Ahlgren, D.J., Verner, I.M.: Fire-Fighting Mobile Robotics and Interdisciplinary Design-Comparative Perspectives. IEEE Transaction on Educations 47(3), 369–376 (2004)
3. Vernerand, I., Waks, S.: Educational features of robot contests: The RoboCup'98survey. Advanced Robotics 14(1), 65–74 (2000)
4. Murphy, R.: Competing for a robotics education. IEEE Robot Automation Magazine 8, 44–55 (2001)
5. Xu, X., Li, S., Ye, Z., Sun, Z.Q.: A survey: RoboCup and the research. In: Proceedings of the 3rd World Congress on Intelligent Control and Automation, vol. 1, pp. 207–211 (2000)
6. Krklješ, D., Nađ, L., Babković, K.: Parametric DC Motor Drive. In: International Symposium on Power Electronics - Ee 2007, Novi Sad, November 7-9, pp. 1–5 (2007) Paper No. T2-1.5
7. Babković, K., Nagy, L., Krklješ, D., Nikolić, M., Borovac, B.: Experiment Platform for Development of Humanoid Robot Foot Control. In: 26th International Conference on Microelectronics - MIEL 2008, Niš, Srbija, Maj 11-14, pp. 459–462 (2008)
8. 6N136 datasheet, http://www.fairchildsemi.com/ds/6N/6N136.pdf
9. BUZ11 datasheet, http://www.radiotechnika.hu/images/BUZ11.pdf
10. IR2110 datasheet, http://www.irf.com/product-info/datasheets/data/ir2110.pdf

A Multi-axis Control Board Implemented via an FPGA

Domenico Longo and Giovanni Muscato

Università degli Studi di Catania, Dipartimento di Ingegneria Elettrica Elettronica
e dei Sistemi – Viale A. Doria 6, 95125 Catania - Italy
{dlongo,gmuscato}@diees.unict.it

Abstract. Most of robotic applications rely on the use of DC motors with quadrature encoder feedback. Typical applications are legged robots or articulated chassis multi-wheeled robots. In these applications system designer must implement multi-axis control systems able to handle an high number of quadrature encoder signals and to generate the same number of PWM signals. Moreover the adopted CPU must be able to execute the same number of control loop algorithms in a time slot of about ten milliseconds. Very few commercial SoC (System on Chip) can handle up to six channels. In this work the implementation of a SoC on FPGA able to handle up to 20 channels within a time slot of 20 ms and up to 100 channels within a time slot of 100 ms is described. In order to demonstrate the effectiveness of the design, the board was used to control a small six wheels outdoor robot.

Keywords: multi-axis control, FPGA, DC motor control, outdoor robot.

1 Introduction

The aim of this project is the development of an embedded system (System on Chip) designed to manage multi-axis platforms where each axis has to be speed or position controlled. Using specific features of the Field Programmable Gate Array (FPGA) devices, the system prototype has been conceived to be able to control an arbitrary number of standalone or related axes (as Dual encoder loop or Virtual Gearing functions). Regardless the majority of the control systems nowadays available on the market, the outstanding project feature is represented by the users capability to customize the number of axis involved in its current application, without being conditioned by the available hardware limitations.

The design of a VHDL entity to quadrature decode encoders signals and one to generate high resolution PWM signals is described. Both use 32-bit wide registers. Inside the Nios II microcontroller (provided by Altera as a softcore), modular software able to provide an arbitrary number of PID control loops, have been implemented and interfaced with developed entities. Thanks to the FPGA modularity and the Nios II programmable features, this System on Chip platform guarantees huge advantages in terms of reliability, compactness and cost, despite of the traditional microcontrollers based solutions. The embedded system is remotely controlled by means of a serial RS232 or Ethernet connection. As an application, the developed system has been used to control a six wheel robot named P6W.

A. Gottscheber, D. Obdržálek, and C. Schmidt (Eds.): EUROBOT 2009, CCIS 82, pp. 88–95, 2010.
© Springer-Verlag Berlin Heidelberg 2010

1.1 The P6W Robot

The P6W (acronym of six wheels prototype) is a small outdoor machine, made of six independent wheels that are linked to six DC motors and powered by batteries. This robot is a small scale version of the Robovolc robot, a robotic systems adopted for the exploration of volcanoes [1]. The purpose of this robot was to allow easy implementation and test of control algorithms directly in laboratory, that were not possible on the full scale system [2]. The robot three main dimensions are 30Lx20Wx20H cm. The robot is a skid-steering one; it is able to move by exploiting the different speed of the wheels of right side from those of the left side. The mechanical structure allows the robot to adapt passively to irregularities of the terrain; it consists of three axes: front, central and rear and has an articulated chassis with four degrees of freedom. Central axis incorporates a metallic chassis used to accommodate the electronics of the system and the joints that allow mating with the other two axes; these joints supplies, the front and rear axis, two degrees of freedom, guaranteeing them the possibility to move vertically and rotate, while taking into account the limits imposed by the structure of the robot.

On each axle, two wheels and two Portescap 17N78 DC motors are associated, which provide a maximum torque of 5.69 mNm; furthermore, in each DC motor a 16 lines quadrature digital encoder and a gearbox, with reduction ratio of 88 : 1, that allows a maximum torque of 0.3 Nm, is incorporated.

An outdoor environments will represent a perfect scenario to test system performance, focusing the attention on the six independent PID control loops supported by suitable hardware, implemented in FPGA, for all the time-critical tasks. Moreover the introduction of particular control functions on-board implemented to create inter-relationships between the six control loops, allowed to test a particular traction control algorithm, that has been previously developed along others DIEES research activities.

2 System Architecture: Hardware Design

The core of the whole system is an FPGA of the Cyclon II Altera family. Using the available logic, a softcore processor (the Nios II, available from Altera) and special I/O peripherals are implemented using VHDL language.

On the Nios II soft processor runs the PID algorithm; this read encoder measurements from the QEIs peripherals and send commands to the motors through the PWMs peripherals. These two special peripherals are represented by different instances of two VHDL entity. The Nios II is connected to these peripherals by means of an internal parallel bus. This approach allows a control system with an high degree of modularity to be developed, as it is possible to define the desired number of channels with very simple operations. In Fig. 1 the described system architecture is shown.

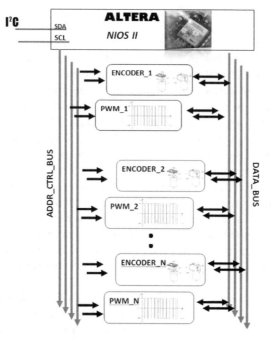

Fig. 1. System architecture

2.1 QEI Block

DC motors are usually associated with optical incremental encoders that generate two quadrature signals with a frequency that is related to the motor speed. In order to obtain speed, position and direction of the motion, a special digital decoding circuit is needed.

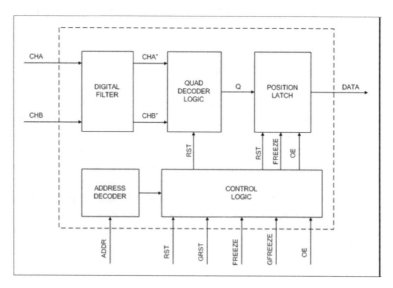

Fig. 2. QEI peripheral block diagram

The block diagram reported in Fig. 2 shows the main components of the QEI (Quadrature Encoder Interface) device. This circuit is implemented in VHDL language and is composed by some blocks that decode and store encoder information and by some other blocks that allows the peripheral to exchange, upon request, information with the Nios II core.

The two encoder channels (CHA and CHB), are filtered before the decoding phase and the signal that is obtained, is subsequently sent to the decoding block. When the control logic requires the counter value, the decoded output is latched on the data bus.

2.2 PWM block

The main PWM device components are shown in the block diagram in Fig. 3. The address logic and the control logic manage the latches access to the DATA_BUS, so that it is possible to memorize the necessary references to make the PWM CORE block able to work properly.

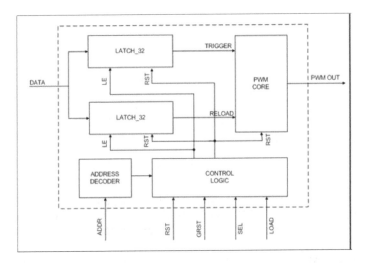

Fig. 3. PWM peripheral block diagram

All the internal registers are 32-bit wide, so the PWM peripheral is able to generate very high resolution PWM signals.

3 System Architecture: Software Design

Also the software developed for the Nios II core has been developed keeping in mind modularity. The code is entirely written in C code and all the functions are defined as vectors, so it is possible to extend the software to control more axis, simply changing few parameters. Moreover the code was separated along different library in order to improve its readability. The resulting software architecture is represented in Fig 4 with the four main software blocks.

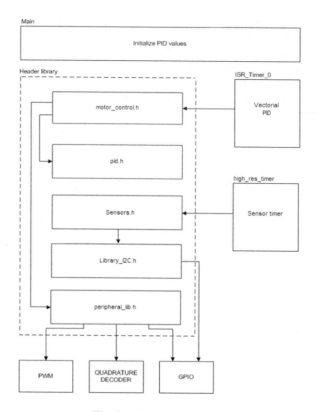

Fig. 4. Software architecture

The "Header Library" is responsible for the management of a specific layer, the "interrupt service routine (ISR)", is responsible for monitoring hardware and software parameters of the system, PID routines is coupled with the timer device, with high priority interrupt.

The block "Main" uses some initialization parameters and implements a global navigation algorithm that allows the robot P6W to move in the presence of obstacles. These libraries can handle also different kind of sensors for navigation purpose. In the same library a particular traction control algorithm is implemented [3].

4 Multi-axis Test on the P6W Robot

The electronic system of the robot mainly consists of:

- ✓ A power board that can provide the appropriate voltages to power amplifiers and to any sensors on board the rover.
- ✓ Three power amplifiers, which are involved in managing the DC motors, in which a current control loop, that enables us to limit the torque for each engine, has been implemented.
- ✓ Six digital incremental encoders, linked to the six motors axles, which provide a measure of the rotational speed of robot wheels.

✓   The FPGA board with a Nios II soft processor, six PWM peripheral and six QEI peripherals, described by VHDL entities.

The purpose of the tests was to verify the effectiveness of the six PID control loops for positioning and for speed regulation. The robot, shown in Fig. 5, was tested also in outdoor environment; the encoder data were acquired on board and sent to a remote PC via an RS232 connection.

References signals from the main CPU to the power amplifiers are sent by means of PWM signals.

Fig. 5. The P6W robot during outdoor testing phase

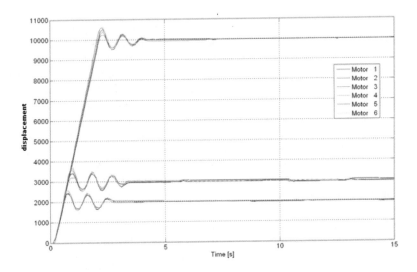

Fig. 6. Six motors position control for different references

In Fig. 6 a graph of the position of each motors is shown for different references. As it can be seen, after a short and small bouncing, each motor is held at the reference position. In Fig. 7 the same measurements for each motor, but referred to speed control, are reported. Also in this case the control system shows very good performance. In both cases, the control algorithm is a standard PID one with anti-windup with a sampling time of 10 ms.

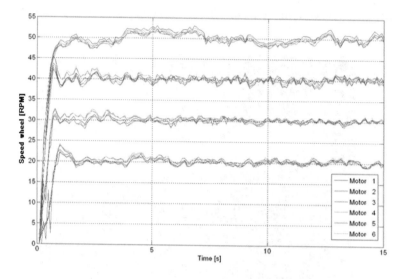

Fig. 7. Six motors speed control for different references

5 Conclusion

In this work a modular System On Chip, implemented on a FPGA, has been described. The system developed has been proved to be modular; this makes it particularly attractive in applications oriented to the control articulated systems, like industrial manipulators characterized by a high number of joints, or applications oriented to control mobile robots. From preliminary tests, with a sample time equal to 100 ms, it has been estimated that the number of controllable axles could be over the hundreds, making it ideal for the control of manipulators or mobile robots of very large complexity. Moreover, also in case of control loops with a sampling period of 10 ms, the high number of axles that can be concurrently controlled is almost twenty, allowing in any case the control of rather complex systems. Possible developments, can be oriented to modify routines monitoring, to reduce the complexity of the code, and to improve efficiency. Other improvements could be obtained from the use of the new functions available in last generation IDE, as the C2H (C to hardware), which allows the processing of portions of code less efficient in blocks dedicated hardware.

In this way the core NIOS II and the peripherals themselves would enjoy the benefits derived from an increase in the operating frequency, in terms of resolution and in terms of performance. Other possible developments could be oriented towards the implementation of a set of libraries for on board generation of trajectories.

Another added value to the whole system, is the opportunity to use in the future a Realtime Operating System as MicroC-OSII or RTAI/uClinux. The compatibility with the Linux kernel, allows in fact the use of the most disparate high-level peripherals such as: Ethernet, Wireless Lan, etc.

Acknowledgment

We would to thank you many people that collaborate to this project through last few years. Among the others, Fabio Rapicavoli, Paola Belluomo, Rino Pappalardo, Massimiliano Minnella, Calabrò Giuseppe, Costa Salvatore Antonio, Foti Sciarampolo Rosario, Mazzara Bologna Giuseppe.

References

1. Muscato, G., Caltabiano, D., Guccione, S., Longo, D., Coltelli, M., Cristaldi, A., Pecora, E., Sacco, V., Sim, P., Virk, G.S., Briole, P., Semerano, A., White, T.: ROBOVOLC: A Robot for volcano exploration – Result of first test campaign. Industrial Robot: An International Journal 30(3), 231–242 (2003)
2. Caltabiano, D., Muscato, G.: A Comparison between different traction methods for a field robot. In: IEEE/RSJ International Conference on Intelligent Robots and Systems, IROS 2002, Lausanne Switzerland, September 30-October4 (2002)
3. Caltabiano, D., Longo, D., Muscato, G.: A New Traction Control Architecture for Planetary Exploration Robots. In: CLAWAR 2005 8th International Conference on Climbing and Walking Robots, London, U.K., September 13-15 (2005)

Robot Localisation in Known Environment Using Monte Carlo Localisation

David Obdržálek, Stanislav Basovník, Pavol Jusko,
Tomáš Petrůšek, and Michal Tuláček

Charles University in Prague, Faculty of Mathematics and Physics
Malostranské náměstí 25, 118 00 Praha 1, Czech Republic
david.obdrzalek@mff.cuni.cz, sbasovnik@gmail.com, pavol.jusko@gmail.com,
petrusek@gmail.com, michal@tulacek.eu

Abstract. In this paper we present our approach to localisation of a robot in a known environment. The decision making and the driving is much harder to be done without the knowledge of the exact position. Therefore we discuss the importance of the localisation and describe several known localising algorithms. Then we concentrate on the one we have chosen for our application and outline the implementation supporting various inputs. Combining of the measurements is also discussed. In addition to well known inputs like odometry and other simple inputs we describe deeper our beaconing system which proved to be very useful.

Keywords: Autonomous robot, Localisation, Monte Carlo Localisation.

1 Introduction

Autonomous robot design is a complex process covering many aspects and containing many decisions. The more autonomous robot we want, the better localisation we need. For example, the localisation of a robot with only one positioning sensor is easy, but the position of the robot can be at stake. On the other hand, as the number of sensors increases, the harder is to estimate the position because of differences in the outputs the sensors provide. One efficient way of dealing with multiple inputs is the use of probabilistic methods. In this paper, we discuss this problem in general, and we present one particular implementation using Monte Carlo localisation (MCL).

The following text is organized as follows: Section 2 gives characterization of the task and presents the specifics of our selected problem. Section 3 gives brief outline of existing localisation techniques. In Section 4, we describe the beaconing system, which provides absolute position information for later use as one of inputs for the localisation process. Section 5 presents Monte Carlo Localisation in general, and Section 6 shows our implementation of it.

2 Characterisation of the Task

The localisation is the key feature of each robot. If a robot does not know its position then its actions are strongly limited. The goal of localisation is to

A. Gottscheber, D. Obdržálek, and C. Schmidt (Eds.): EUROBOT 2009, CCIS 82, pp. 96–106, 2010.

determine the most probable position of the robot based on known data. The localisation algorithm itself should be universal and independent on the actual purpose of the robot. Then, the only variable part of a localisation procedure is the set of sensors.

We will use data from robot's sensors to compute its exact position. For example, we can use odometry sensors, contactless detectors of obstacles, beacons for direction and/or distance measuring, bumpers, camera and others. All these sensors provide data with different accuracy so we need to handle the data with different weights. Localisation algorithms should not depend on a specific type of sensor or a limited list of them. The goal is to develop a universal solution that works with different sensors (or better said with as many different sensors as possible). Information retrieved from these sensors is consequently used as input data for localisation algorithm.

In this article we will present the algorithm on the robot which is designed to participate in the Eurobot autonomous robot contest [1]. The Eurobot contest rules define the following properties of the task:

- Localisation is done in a known indoor environment
- Important areas of environment are highlighted
- The robot starts placed at known position
- The task is limited by time - there are only 90 seconds available
- The environment is relatively small (2×3 meters)
- There is one opponent robot which competes with our robot
- The environment contains several game elements
- The robot has to find important areas of the environment
- There can be placed localisation beacons at predefined places
- The robot has limited height and circumference.
- Because of its size, the robot has only a limited computing capacity

This list is quite wide. Obviously, it affects the development of robot hardware and software, but it should be also noted that the algorithm itself is not that much application dependent and can be used for other applications in different conditions too.

The localisation must be done with reasonable precision. If the robot computes wrong position it could accidentally hit and destroy built structures or miss its goal. This would surely lead to losing of the game. Also, everything has to be computed almost in real-time. In Eurobot game, the robot has 90 seconds to locate game elements on the playing field or stored in the dispensers, collect the elements, drive to the target area and finally build there a defined structure according to the rules. Throughout this time it therefore needs to know in each moment where exactly it is and avoid collisions with obstacles and the opponent robot.

As can be seen, GPS-based sensors cannot be used – the indoor environment and small working area makes this sensor unusable. For such environment, other solution must be used. Even we can assume that there is coherent light with a few local discrepancies which helps to use visual systems, we need to expect interferences in sensor readings - ultrasonic signals reflect in corners, infrared sensors are affected by strong lights, and radio sensors are disturbed by local radio communication. On the other hand we can expect that the robot will not meet unexpected obstacles (except playing elements or opponent robot).

3   Localisation Algorithms

The area of autonomous robot localisation is well researched (see e.g. [2]), and several ways can be used to solve the localisation problem. Therefore we do not try to invent a new algorithm. Instead, we will outline some existing localisation algorithms and discuss some of their implementation details, together with technical problems we have met.

For localisation based on various input values we can choose from many algorithms. Here we will present some of them:

Kalman filter [3,4] the method which generalizes the floating mean. It can handle noisy data so it is suitable for processing the data from less precise sensors. However, the model must be described with the expected value and variability which is often too difficult constraint.

Markov localisation based on grid [5] this method resolves problem of the Kalman filter, that must know expected value and variance of input data. This algorithm splits the area to the grid of proper size. However this effort needs large operational memory and computing power.

Monte-Carlo localisation [6] method that can be easily implemented. It can represent multimodal distribution and thus localize the robot globally. Moreover it can process inputs from many sensors with different accuracy.

For our implementation we have chosen the Monte-Carlo localisation for its simplicity and possibility of using more sensors.

4   Beacons

In this section, we present our design of a sensor set which provides relatively good information about the robot position and in cooperation with other sensors it helps to create a robust localisation system.

The main idea of this beacon system is to mount several beacons around the working area of the robot and let the robot measure the distance to these beacons. Then, the robot will be able to estimate its position because the beacons position is known.

4.1   Physical Principle

Our absolute positioning system works on a simple principle

$$l = c \cdot dt \tag{1}$$

where l is the distance, c is the speed and dt is the time of travel. In our system, we measure the time the signal travels from the transmitter at the beacon to the receiver mounted on the robot. Of course this works only if the speed is constant. This condition is met as we are using ultrasonic waves. Since the speed of these sound waves is known, we have to measure the time difference, from which the distance may be calculated using Equation (1). To correctly measure the travelling time, we synchronize the system by using infrared light. Similar systems based on Equation (1) (although using different time synchronisation) have been developed; for examples see e.g. [7,8].

4.2 Hardware

The transmitting system consists of three separate stationary beacons connected by a cable. When the system receives signals from the three beacons and calculates the three distances, it can theoretically determine its position by using triangulation. In praxis, such simplest form does not fully work. The robot may move between receiving signals from all the beacons, some signals may not be received or they may provide incorrect information. Even that, good estimation may be acquired as will be further discussed in Section 6.

The signal is sent one-way only, the receivers do not communicate with the transmitters. Therefore, there can be multiple independent identical receivers, which are able to determine its individual positions. These receivers may be then used for localizing more objects and if the information is passed to a single centre, it may be of substantial help.

Now we will focus on each part of the system closer.

Transmitters work in the following way.

1. Send timing information from one beacon (infrared)
2. Wait for a defined period of time
3. Send distance measuring information from the beacon (ultrasonic)
4. Wait a defined period of time
5. Repeat steps 1-4 for all other beacons.

Since we want to measure time difference between the signal being sent and it being received, we need to have synchronized clocks. This is done in step 1 by using infrared modulated signals. Besides that, the transmitted information contains additional information about the beacon which is going to transmit sound waves.

Sound waves are transmitted as ultrasonic waves, and are always transmitted only from one beacon at a time. The transmitted signal contains also the identification of the source beacon.

Receiver. waits for infrared timing information. When it is received, the receiver resynchronizes its internal timer and generates a message. These messages are transported to the localisation unit by serial line (RS-232), but any other transport layer may be used (e.g. Bluetooth). Every message contains time stamp, information that synchronisation occurred, and the information about beacon which is going to transmit ultrasonic information in this time step. Upon reception of the timing information, the receiver switches its state to wait for ultrasonic signal. When correct ultrasonic information arrives, the receiver generates similar message as is the message after IR reception, but containing time stamp for ultrasonic reception and beacon identification transmitted in the ultrasonic data. The difference in these two timestamps is related to distance (only a constant needs to be subtracted, because the two signals are not transmitted exactly at the same time). Since each beacon identifies itself in both infrared and ultrasonic transmissions, the probability of mismatch is reduced.

When the infrared information is not received, a message is generated saying the synchronisation did not occur and the timestamp is generated from previously synchronised internal clock. When the ultrasonic information is not received, localisation unit is notified that nothing was received.

The situation after three successfully received ultrasonic signals with synchronised clock can be seen in Figure 1.

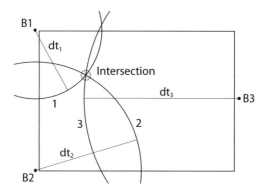

Fig. 1. Beaconing system

5   General Description of MCL

In this section we will briefly introduce Monte Carlo Localisation (MCL). This algorithm meets all the requirements mentioned in problem statement section earlier in this paper. It is a well defined and researched algorithm (see e.g. [6]) and it is also well established in many applications (see e.g. [9,10,11,12]).

Monte Carlo Localisation maintains a list of robot's possible states or positions. Each state is weighted by its probability of correspondence with the actual state of the robot. In the most common implementation, the state represents the coordinates in 2D Cartesian space and the heading direction of the robot. It may be of course easily extended to 3D space and/or contain more information depicting the robot state. All these possible states compose the so called probability cloud.

The Monte Carlo Localisation algorithm consists of three phases: the prediction, measurement, and resampling phase.

During the prediction phase, a new value for each item of the cloud is computed, resulting in a new probability cloud. To simulate various inaccuracies that appear in a real hardware, random noise is added to each position in the prediction phase. This is very useful. For example: If the wheels were slipping and no random noise was added, the probability cloud would travel faster than the real hardware.

During the measurement phase, data from real sensors are processed to adjust probability of the positions in the cloud. The probability of samples with lesser likelihood (according to sensors) is lowered and vice versa. For example, when the sensors show the robot orientation is northways, weight for samples representing other orientations is lowered.

The last phase - resampling - manages size and correctness of the cloud. Positions with probability lower than a given threshold are removed from the cloud. To keep the number of positions constant, new positions are added. These new positions are derived from and placed around the existing positions with high probability.

More formally, our goal is to determine robot's state in step k, presuming the original state and all the measurements $M^k = \{m_i, i = 1..k\}$ are known. Robot's state is given by a vector $x = [x, y, \alpha]$, where x and y is the robot position and α is its heading.

1. During the prediction phase, the probability density $p\left(x_k \mid M^k\right)$ enumeration for the step k takes place. It is based only on presumed movement of the robot without any input from real sensors. Therefore, for any known command u_{k-1} given to the robot, we have

$$p\left(x_k \mid M^{k-1}\right) = \int p\left(x_k \mid x_{k-1}, u_{k-1}\right) p\left(x_{k-1} \mid M^{k-1}\right) dx_{k-1} \qquad (2)$$

2. In the measurement phase, we will compute the final value of probability density for actual step k. To do so we will use data from sensors expressed as probability $p\left(m_k \mid x_k\right)$, where m_k is the actual state and x_k is the assumed position. Finally, probability density in step k is described by the following equation:

$$p\left(x_k \mid M^k\right) = \frac{p\left(m_k \mid x_k\right) p\left(x_k \mid M^{k-1}\right)}{p\left(m_k \mid M^{k-1}\right)} \qquad (3)$$

Initialization:

- If the robot's position is known, then for its state x we have: $p\left(x \mid M^0\right) = 1$, for all other states $y \neq x$ $p\left(y \mid M^0\right) = 0$.
- If the robot's position is not known, the probability of all positions is the same. Therefore $p\left(x \mid M^0\right)$ must be set for all x so that

$$\int p\left(x \mid M^0\right) dx = 1 \qquad (4)$$

One of the most important features of this method is its ability to process data from more than one source. Every sensor participates on computing the probability for the given state. For example, if we have a compass sensor and it reads that the robot is heading to the north, we can lower the probability of differently oriented samples. If robot's bumper signalizes collision, there is a high probability for the robot to be near a wall or another obstacle. It is therefore possible to discard the part of the probability cloud which lies in an open space.

Our implementation of the MCL algorithm is described in more detail in the following section.

6   MCL Implementation

Our implementation of MCL is based on previous work on the Logion robot which participated in Eurobot 2008 contest (see [13]), which has been further developed and extended. In the following paragraphs, we emphasize several implementation aspects we consider as important for the successful result.

6.1   Sensors Processing

In our implementation, we divide the inputs coming from sensors in two categories:

- Advancing inputs
- Checking inputs

Our system contains two interfaces for these two types of inputs. The device or its abstraction in Hardware Abstraction Layer implements the corresponding interface based on its type, so the MCL core can use it as its input. The MCL core calls each device when it has new data, and the work with the samples is done by each device separately. This keeps the main code easier to read, simpler, and input independent. Also, the device itself knows the best how to interpret the raw data it measures.

The level of reliability can be specified for each input device. Then, the samples are adjusted by the devices with respect to their configured credibilities. For example: two sets of odometry encoders, one pair on driven wheels and one pair on dedicated wheels, have different accuracy because the driven wheels may slip on the surface when too much power is used. Then, the credibility of driven wheels encoders will be set lower than the credibility of the undriven sensor wheels. In addition, setting the reliability level helps to deal with different frequencies of data sampling.

6.2   Advancing Inputs

This input type is used for moving the samples. Such input could be for example odometry (processing of wheel encoders). The information provided by these kind of inputs applied to samples is blurred by randomly generated noise as described earlier. After advancing the samples, boundary conditions are checked.

As a result the probability of samples representing impossible positions is decreased. If there are several advancing input devices, these devices must be read all at once and the result advancing information is computed as a weighted average of all inputs according to their reliabilities.

6.3 Checking Inputs

Checking inputs are not affecting the position of the samples. Instead, they are just adjusting their probability (also called sample weights). The reason for this is that inputs of this type do not provide relative difference from the last measurement, but absolute position information. This also does not need to be one exact point, but an area or position probability distribution, which fits perfectly to the Monte Carlo Localisation algorithm. All checking inputs are processed separately; we regulate them only by setting their reliability levels.

Our robot uses these checking inputs:

- compass - checks the direction of samples
- beacons - checks the distance from stationary beacons
- bumpers - checks collisions with playing field borders and other objects
- IR distance sensors - checks distance to borders and obstacles

6.4 Position Estimation

It is expected that the MCL outputs the estimation of robot position. Because of its nature and implementation, the result position can be computed from samples at any time. This estimation is very simple, just computing the weighted average of all samples. In addition we can determine the overall reliability of this estimation.

6.5 I am Entirely Lost

MCL can also determine the robot position from scratch if it has some absolute sensors. The only change to the algorithm itself is reinitializing of the sample cloud. At the beginning of localisation (when the robot is lost) samples are spread uniformly all over the playing field as described in Section 5. The sensors providing absolute positioning information lower the weight of misplaced samples and new samples are placed in regions with higher probability (see Figure 2). This is repeated until sufficiently reliable position estimation of the robot is reached.

6.6 Beacons and MCL

As described earlier, our beacon system consists of three transmitting and one receiving beacons. The information is passed from the beacon system to the main computing unit via messages containing beacon id (i.e. transmitter identification) and time difference between the infrared and ultrasonic transmissions. For more hardware details see full description in Section 4.

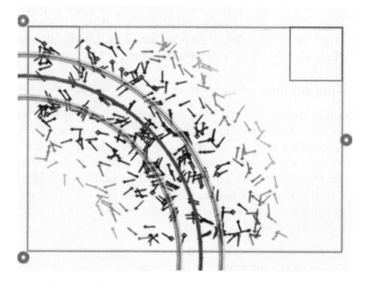

Fig. 2. MCL after processing one beacon input: The circular belt marks the input from the bottom left beacon. The "pins" represent oriented MCL samples; sample probability is proportional to their darkness.

There are two reasons why each message contains the time difference (delta) instead of the calculated distance: computational power of the micro controller and the degree of robustness. The main computing unit is more powerful than the receiving beacon, so we let the beacon do less work and we even benefit from this decision. We considered deltas to be the perfect raw data for our purpose - distance measurement. The computation is done in the main computing unit which controls all the other devices and is highly configurable. It means that all the parameters of the equation for distance calculation can be changed easily without the need of changing the beacons hardware or device firmware. It even allows us to calculate or adjust the parameters on the flight if distance information is provided based on external measurement. The main calculation is described by Eq. (1).

The configuration of the main computing unit contains not only the important constants for the equation, but also the positions of the transmitting beacons. As we know the distance and the beacon id, we can increase the weights of the MCL samples in the circular belt formed by these two values and a range constant. MCL samples far from the belt are penalized (see Figure 2).

This approach is much better and more robust than just waiting for intersections and then computing the robot position using simple triangulation. These intersections may not happen very often because of the time gap between individual beacon transmissions (especially when the robot is moving fast). At the same time, it is good to implement different weighting for the samples on a belt, near an intersection of two belts and near the intersection of all three belts.

6.7   Camera

The idea of using camera for absolute positioning seemed very hard at the first time. Later, when we had the modular MCL implementation finished, we realized there is a great opportunity to use the information we get from the camera while looking for the playing elements positioned at predefined places of the playing area. Now, we can compare the playing element positions (acquired from the camera) with their fixed positions (defined by the Eurobot contest rules) and adjust the weight of the MCL samples to merge the two positions.

6.8   Gyroscope

In the early stages of robot design, we proposed to use a compass as one of the input sensors. However, using a compass in a small indoor competition is not a very good idea, because its precision can be degraded by influence of many factors (e.g. huge metallic cupboard, electromagnetism, steel concrete walls or metal structure building). Using a gyroscope instead of a compass would be much more efficient for our purposes, because gyroscope works completely independently and the influence of the environment is minimal. The only problem is the placement of the gyroscope itself, because it should be placed in the rotational axis of the robot.

6.9   Performance

Apparently, the processing of a large number of MCL samples may have impact on performance of the whole localisation system. In general, the more samples we take into computation, the more precise the localisation is, but the slower the computation is. There is an easy way to adjust the precision to speed ratio by setting the number of samples in the MCL cloud. In our project, we have achieved acceptable precision and speed using 400 samples.

7   Conclusion

The advantages of using Monte Carlo Localisation instead of relying on straightforward sensor measurings are great. Now we can use different sensing devices and combine them with respect to their parameters, precision and credibility.

We have developed a modular system for robot localisation which allows easy extension by different kinds of modules. Our implementation allows us to add more facilities with almost no or just minimal work effort and with no changes to the MCL itself at all, while increasing the precision of the resulting position.

In our paper we have described the advantages of the Monte Carlo Localisation compared to other methods of position estimation and how we benefit of it in our implementation. The modular design of sensor processing and MCL position estimation is described, together with practical comments on different sensor types. The created system will be used for Eurobot 2009 contest edition but its design allows using it for other purposes too (which we are already planning).

Acknowledgement

This work has been partially supported by The Charles University Grant Agency and by the National Programme of Research, Information Society Project number 1ET100300419.

References

1. Eurobot: Eurobot autonomous robot contest (2009), http://www.eurobot.org
2. Thrun, S.: Robotic mapping: a survey. In: Exploring Artificial Intelligence in the New Millennium, pp. 1–35. Morgan Kaufmann Publishers Inc., San Francisco (2003)
3. Negenborn, R.: Robot localisation and kalman filters: On finding your position in a noisy world. Master's thesis, Utrecht University (2003)
4. Welch, G., Bishop, G.: An introduction to the kalman filter. Technical Report TR 95-041, University of North Carolina at Chapel Hill (2004)
5. Burgard, W., Derr, A., Fox, D., Cremers, A.B.: Integrating global position estimation and position tracking for mobile robots: The dynamic markov localization approach. In: Proc. of the IEEE/RSJ International Conference on Intelligent Robots and Systems, IROS (1998)
6. Dellaert, F., Fox, D., Burgard, W., Thrun, S.: Monte carlo localization for mobile robots. In: Proc. of the IEEE International Conference on Robotics & Automation, ICRA'99 (1998)
7. Yi, S.y.: Global ultrasonic system with selective activation for autonomous navigation of an indoor mobile robot. Robotica 26(3), 277–283 (2008)
8. Dazhai, L., Fu, F.H., Wei, W.: Ultrasonic based autonomous docking on plane for mobile robot. In: IEEE International Conference on Automation and Logistics (ICAL 2008), pp. 1396–1401 (2008)
9. Menegatti, E., Zoccarato, M., Pagello, E., Ishiguro, H.: Image-based monte-carlo localisation with omnidirectional images. Robotics and Autonomous Systems 48, 17–30 (2004)
10. Hähnel, D., Burgard, W.: Mapping and localization with rfid technology. In: Proc. of the IEEE International Conference on Robotics & Automation (ICRA'05), pp. 1015–1020 (2004)
11. Wulf, O., Khalaf-Allah, M., Wagner, B.: Using 3d data for monte carlo localization in complex indoor environments. In: 2nd Bi-Annual European Conference on Mobile Robots (ECMR'05), pp. 170–175 (2005)
12. Lenser, S., Veloso, M.: Sensor resetting localization for poorly modelled mobile robots. In: Proc. of the IEEE International Conference on Robotics & Automation, ICRA'00 (2000)
13. Mikulik, A., Obdrzalek, D., Petrusek, T., Basovnik, S., Dekar, M., Jusko, P., Pechal, R., Pitak, R.: Logion - a robot which collects rocks. In: Proceedings of the EUROBOT Conference 2008, pp. 276–287 (2008)

Detecting Scene Elements
Using Maximally Stable Colour Regions

David Obdržálek, Stanislav Basovník, Lukáš Mach, and Andrej Mikulík

Charles University in Prague, Faculty of Mathematics and Physics
Malostranské náměstí 25, 118 00 Praha 1, Czech Republic
david.obdrzalek@mff.cuni.cz, sbasovnik@gmail.com,
lukas.mach@gmail.com, andrej.mikulik@gmail.com

Abstract. Image processing for autonomous robots is nowadays very popular. In our paper, we show a method how to extract information from a camera attached on a robot to acquire locations of targets the robot is looking for. We apply maximally stable colour regions (a method originally used for image matching) to obtain an initial set of candidate regions. This set is then filtered using application specific filters to find only the regions that correspond to scene elements of interest. The presented method has been applied in practice and performs well even under varying illumination conditions since it does not rely heavily on manually specified colour thresholds. Furthermore, no colour calibration is needed.

Keywords: Autonomous robot, Maximally Stable Colour Regions.

1 Introduction

Autonomous robots often use cameras as their primary source of information about their surroundings. In this work, we describe a computer vision system capable of detecting scene elements of interest, which we used for an autonomous robot. The main goals we try to achieve are robustness of the method and computational efficiency.

The core of our vision system are *Maximally Stable Extremal Regions*, or MSERs, introduced by Matas et. al (see [1]) for gray-scale images and later extended to colour as *Maximally Stable Colour Regions*, or MSCR (see [2]). Details about MSER and MSCR principles are given in Sections 2 and 3, respectively.

The main usage of MSER detection is for wide-baseline image matching mainly because of its affine covariance and high repeatability. To match two images of the same scene (taken from different viewpoints), MSERs are extracted from both images and then appropriately described using (usually affinely invariant) descriptor (see [3,4]). Because MSER extraction is highly repeatable, the majority of the regions should be detected in both images. If the descriptor is truly affinely invariant, identical regions should have the same (or similar) descriptors even though they are seen from different viewpoints (assuming the regions correspond to small planar patches in the scene). Then, the matching can be done using nearest neighbour search of the descriptors.

A. Gottscheber, D. Obdržálek, and C. Schmidt (Eds.): EUROBOT 2009, CCIS 82, pp. 107–115, 2010.
© Springer-Verlag Berlin Heidelberg 2010

In our system, MSCRs are not used for matching but for object detection. The system operates in the following steps:

- Detect large number of contrasting regions in the image.
- Classify detected regions and decide which correspond to elements of interest.
- Localize detected elements and pass this information to other components of robot's software.

MSER and MSCR algorithms often return large number of (possibly overlapping) regions. We therefore introduce our classification algorithm which rejects regions with small probability of corresponding to a scene element of interest. The relative position of the scene element is then determined using standard algorithms from computer vision and projective geometry [5].

Typical input image and output in the form of list of detected objects can be seen in Figure 1.

Localization results	
object 1	[-0.74, 0.72, 0.01]
object 2	[0.84, 1.97, 0.05]
object 3	[-0.81, 2.20, -0.01]
object 4	[0.23, 3.21, 0.04]

Fig. 1. Input image, detected regions, and final output table – triangulated coordinates of detected objects

The following text is structured as follows: We first briefly describe the MSER (Section 2) and MSCR (Section 3) algorithms. In Section 4, we present our filtering system which processes the regions and outputs locations of detected objects. Section 5 discusses the overall efficiency of the proposed algorithm.

2   MSER

In this section, we describe the MSER algorithm as a basis for our region detection.

The MSER detection uses a watershedding process that can be described in the following way:

The gray-scale image is represented by function $I : \Omega \rightarrow [0..255]$, where $\Omega = [1..W] \times [1..H]$ is the set of all image coordinates. We choose an intensity threshold $t \in [0..255]$ and divide the set of pixels into two groups B (black) and W (white).

$$B := \left\{ \mathbf{x} \in \Omega^2 : I(x) < t \right\}$$

$$W := \Omega^2 \setminus B$$

When changing the threshold from maximum to minimum intensity, the cardinality of the two sets changes. In the first step, all pixel positions will be contained in B and W is empty (we see completely black image). As the threshold t is lowered, white spots start to appear and grow larger. White regions grow and eventually all merge when the threshold reaches near minimum intensity and the whole image will be white (all pixels are in W and B is empty). Figure 2 demonstrates the evolution process with different threshold levels.

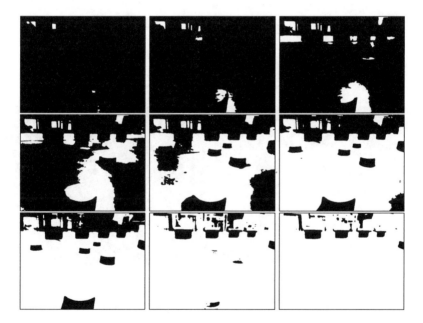

Fig. 2. MSER evolution of the input image shown in Figure 1. Results of 9 different thresholding levels are displayed, each time for lower intensity threshold t.

Connected components in these images (white spots and black spots) are called *extremal regions*. Maximally stable regions are those that have changed in size only a little across at least several intensity threshold levels. The number of levels needed is a parameter of the algorithm.

3 MSCR

In this section, we outline the MSCR method as extension of MSER from gray-scale to colour images (see [2]).

In the following text, we assume the image to be in RGB colour space, but it can be easily seen the MSCR method can work with other colour spaces too. To detect MSCRs, we take the image function $I : \Omega \to \mathbb{R}^3$. Thus, the image function I assigns a colour (RGB channel values) to all pixel positions in the given image. We also define graph G, where the vertices are all image pixels, and the edge set E is defined as follows (note that $\mathbf{x}$, $\mathbf{y}$ are 2-dimensional vectors):

$$E := \big\{ \{\mathbf{x}, \mathbf{y}\} \in \Omega^2 : |\mathbf{x} - \mathbf{y}| = 1 \big\}$$

where $|\mathbf{x} - \mathbf{y}|$ is a Euclidean distance of pixel coordinates $\mathbf{x}$ and $\mathbf{y}$ (other metrics, e.g. Manhattan distance, can be considered too). Edges in the graph connect neighbouring pixels in the image. Every edge is assigned with the weight $g(\mathbf{x}, \mathbf{y})$ that measures the colour difference between the neighbouring $\mathbf{x}$ and $\mathbf{y}$ pixels. In accordance with [2], we use the *Chi squared measure* to calculate the value of g:

$$g^2(\mathbf{x}, \mathbf{y}) = \sum_{k=1}^{3} \frac{(I_k(\mathbf{x}) - I_k(\mathbf{y}))^2}{I_k(\mathbf{x}) + I_k(\mathbf{y})}$$

where $I_k(\mathbf{x})$ denotes the value of the k-th colour channel of the pixel $\mathbf{x}$.

(a) scale=0.25   (b)

Fig. 3. Blurred input image and first stage of MSCR evolution

We then consider series of subgraphs $E_t \subseteq E$, where the set E_t contains only edges with weight $\leq t$. The connected components of E_t will be referred to as regions. In the MSCR algorithm, we start with $E_t, t = 0$ and then gradually increase t. As we do this, new edges appear in the subgraph E_t and regions start to grow and merge. MSCR regions are those regions that are stable (i.e., nearly unchanged in size) across several thresholding levels, similarly to MSER algorithm.

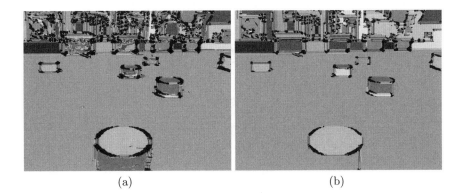

(a)   (b)

Fig. 4. Evolution of regions with increasing threshold

For an example of detected MSCR regions, refer to Figures 3 and 4: Figure 3(a) shows the input image (after it is blurred using Gaussian kernel as is usually done in image segmentation to handle the noise). Figure 3(b) shows regions of the graph with edges E_0 (represented in false colours: different components of this graph have different colours, trivial isolated 1 pixel components are black). Figures 4(a) and 4(b) show two further stages of the computation – as we increase the threshold t, we can see the homogeneous parts of the image merge and form regions. We can see that the contours of important scene elements can be clearly distinguished on the latter two images.

4   Filtering Regions

This section shows how the set of regions detected by MSCR is filtered so that only interesting regions are kept and regions without the importance to the application are discarded.

Using MSER and MSCR, we retrieve quite large number of image regions (see Figure 5), of which only a few is of any importance. Therefore, this set of regions has to be filtered to discard all regions of no interest. This part is application specific and depends on the appearance of the objects that are being detected. In our testcase, the robot operates on a relatively small space with flat single coloured surface and it interacts with scene elements of two colours, red and green. So, the scene elements have contrasting colours to the background, which is a standard assumption for successful object detection in a coloured image.

In the following paragraphs, we show the individual filters that were successively applied on the original set of regions. The result is a list of objects, which is passed for further processing in the robot planning algorithm. Figure 5 shows the input image and the first set of regions, which is to be filtered, Figure 6 shows the situation after each step.

Fig. 5. Input picture and all detected regions denoted by black contour

4.1 Discarding Regions Touching Image Border

A useful heuristic to reject regions detected outside of the working area is to discard regions touching the image border. Of course, this way we may also reject contours that really correspond to important scene element. However, since reliable detection of objects that are only partially visible is considerably harder, we have decided to reject them without substantial loss. Figure 6(b) shows the regions which remain after dropping regions touching the border. In this particular case, no loss of interesting regions occurred.

4.2 Shape Classification

After the contour is detected, the polygonal shape is then simplified by using the classic Douglas-Peucker algorithm (see e.g. [6]). If the resulting polygon is too large or too small (above or below specified parameters), it is rejected. The lower and upper limits are set according to expected size of objects the robot should detect.

Also, in our case the objects of interest all lie on the ground and their shape contours contains parallel lines (e.g. a cylinder). Therefore, the detected contours should contain long segments that are either horizontally or vertically aligned with the image coordinate system. If such segment cannot be found, the region is rejected. Figure 6(c) shows the set of regions after applying this filter.

4.3 Statistical Moments (Covariance Matrix)

Contours corresponding to the top parts of objects can often be nicely and reliably isolated and detected. The column elements, which are used in Eurobot 2009, have circular shape and therefore the image of the top part is an ellipse (for details about Eurobot autonomous robot contest, see [7]). We therefore take special care to distinguish elliptical regions from all others. To do this, we calculate first, second and third statistical moments of the region. Since ellipse is a second-order curve, its third statistical moment should ideally be zero (in real case, close to zero). In Figure 6(d), the regions satisfying this criterion are highlighted.

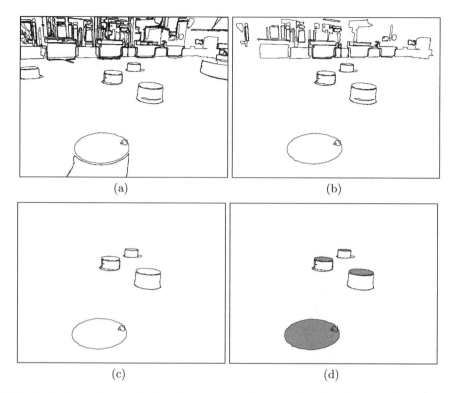

(a)   (b)

(c)   (d)

Fig. 6. Regions after each filtering step. In Figure 6(d), the regions satisfying the covariance matrix condition are filled.

In addition, the first statistical moment is preserved for later use – to localize the centre of gravity of the column element.

4.4 Colour Classification

Finally, it is necessary to decide whether the detected object is red, green or has some other colour, because in the Eurobot 2009 game, only elements with colour assigned to the team may be handled. When determining the colour of the object, the average colour of the region is considered. Since the measured colour depends heavily on the illumination, this check is the last one in our pipeline and we reject regions based on their colour only in cases where the colour significantly differs from red or green.

4.5 Position Calculation

After the regions are filtered and their colour is classified, we calculate their position on the playground. It is possible to calculate the object position because we know the camera parameters (from its calibration) and we also know that the objects lie on the ground. We then save this information into appropriate data structures and pass it to other components of robot's software to be used e.g. for navigation.

4.6   Final Output

After the filtering, the remaining regions are claimed to represent real objects on the playing field. Figure 1 shows the resulting table of objects. For further processing, only these coordinates are sent out. This is a very little amount of data and at the same time, it is a very good input for the central "brain" of the robot, which uses this information for example as data for trajectory planning process.

5   Computational Efficiency

Robots often have limited computational power which restricts the use of computationally intensive algorithms. In this section, we discuss the computational complexity of the method.

The (greyscale) MSER algorithm first sorts individual pixels according to their intensity. Since the pixel intensity is an integer from the interval $[0..255]$, this can be done in $O(n)$ time using radix sort. During the evolution process, regions are merged as new pixels get above the decreasing threshold t. To do this, the regions must be stored in memory in appropriate data structures and a fast operation for merging two regions has to be implemented. This is straightforward application of the union-find algorithm with time complexity of $O(n\alpha(n))$, where $\alpha(n)$ is the inverse Ackermann function – an extremely slowly growing function. Therefore, this part does not bring in real cases significant time demands.

The MSCR variant differs from MSER in two respects. Individual pixels are not considered; instead the edges between neighbouring pixels are taken into account. This increases the number of items by factor of 2. Also, the colour difference between neighbouring pixels (in *Chi squared meassure*) is a real number and the sorting part thus takes $O(n \log(n))$ time.

Once the regions are detected, most of them are quickly rejected based on their size (too small or too large) or position (near the image border). Only limited amount of regions must be processed using more complex filters. In practice, this does not significantly increase the computational time: in our example at Figures 5 and 6, only 72 regions remained after application of the first filter.

6   Conclusion

In this paper we have shown how MSER and MSCR algorithms can be used for detection of objects in an image in one practical application. The resulting method is robust in respect to illumination changes as it uses classification by colour only as its last step.

As a final result of our tests, we are able to process 5-10 images (320×240 px) per second on a typical netbook computer without any speed optimizations.

This could be further improved by e.g. using CPU specific instructions such as SSE, but even this speed is sufficient for our purpose – to provide locations of objects which the robot has to handle.

Acknowledgement

This work has been partially supported by The Charles University Grant Agency and by the National Programme of Research, Information Society Project number 1ET100300419.

References

1. Matas, J., Chum, O., Urban, M., Pajdla, T.: Robust wide baseline stereo from maximally stable extremal regions. In: Proceedings of the British Machine Vision Conference 2002 (BMVC'02), pp. 384–393 (2002)
2. Forssén, P.-E.: Maximally stable colour regions for recognition and matching. In: IEEE Conference on Computer Vision and Pattern Recognition, CVPR'07 (2007)
3. Matas, J., Obdrzalek, S., Chum, O.: Local affine frames for wide-baseline stereo. In: Proceedings of the 16th International Conference on Pattern Recognition, ICPR'02 (2002)
4. Forssén, P.E., Lowe, D.: Shape descriptors for maximally stable extremal regions. In: IEEE International Conference on Computer Vision, ICCV'07 (2007)
5. Hartley, R.I., Zisserman, A.: Multiple View Geometry in Computer Vision. Cambridge University Press, Cambridge (2004) ISBN: 0521540518
6. Douglas, D., Peucker, T.: Algorithms for the reduction of the number of points required to represent a digitized line or its caricature. Canadian Cartographer 10, 112–122 (1973)
7. Eurobot: Eurobot autonomous robot contest (2009), http://www.eurobot.org

Ultrasonic Localization of Mobile Robot Using Active Beacons and Code Correlation

Marek Peca

Department of Control Engineering, Faculty of Electrical Engineering,
Czech Technical University in Prague, Czech Republic
pecam1@fel.cvut.cz
http://dce.felk.cvut.cz/

Abstract. Ultrasonic localization system for planar mobile robot inside of a restricted field is presented. System is based on stationary active beacons and measurement of distances between beacons and robot, using cross-correlation of pseudorandom binary sequences (PRBSes). Due to high demand of dynamic reserve imposed by range ratio in our specific task, both code- as well as frequency-divided media access has been utilized. For the same reason, 1-bit signal quantization has been abandoned in favor of higher resolution in receiver analog-to-digital conversion. Finally, dynamic estimation of the position is recommended over analytic calculation. The final solution uses the extended Kalman filter (EKF), equipped with erroneous measurement detection, initial state computation, and recovery from being lost. EKF also performs data-fusion with odometry measurement. Unlike the approach in majority of works on mobile robot localization, a model, actuated solely by additive process noise, is presented for the data-fusion. It offers estimation of heading angle, and remains locally observable. Simplistic double integrator model of motion dynamics is described, and the importance of clock dynamics is emphasized.

1 Design Considerations

1.1 Objectives

The system has been designed for a planar wheeled robot navigation inside a restricted, rectangular 3×2.1 m playground according to the EUROBOT competition rules [1]. Around the playground, there may be placed at least three stationary beacons. They may be interconnected together by a wire.

Our robot is a conventional two-wheel vehicle. Its floor projection is a rectangle 258×290 mm. Beacon counterpart is a "sensor area" on top of the robot in rectangle center. Beacon to sensor area distance ranges between $0.19 \ldots 3.6$ m. Rough estimate of maximal robot velocity is $2\,\mathrm{ms}^{-1}$.

1.2 System Fundamentals

The system is based on wave propagation and signal time-of-flight measurement. Ultrasound has been selected for its slow propagation speed, compared to

A. Gottscheber, D. Obdržálek, and C. Schmidt (Eds.): EUROBOT 2009, CCIS 82, pp. 116–130, 2010.

electromagnetic waves. Therefore, cutting-edge high-frequency hardware is not required. Our system is characterized by following fundamentals:

- three static transmitters (beacons), one mobile receiver on the robot
- time displacement evaluation based on pseudorandom binary sequences (PR-BSes) correlation
- continuous simultaneous transmission of modulated PRBSes
- both code-divided multiple acces (CDMA) and rudimentary frequency-divided multiple access (FDMA) to provide sufficient dynamic reserve
- Doppler effect ignored

The beacons can be either active (transmitting), passive (receiving), or reflective (returning the signal to the robot). Measurement data need to be collected at the robot, what lead to choice of active beacons.

Distance measurement systems based on wave propagation in general use either short pulses [2], or pseudorandom signals, due to desired correlation properties. Auto-correlation of signals of both types approaches single, narrow peak (Sec. 2.1). Continuous transmission of pseudorandom signals has been favoured over the pulses interleaved by long pauses, due to better utilization of communication channel capacity, and rosistance against short-time disturbances. Only binary pseudorandom signals (PRBSes) have been used, mainly for their easy generation and amplification in simple hardware (Sec. 2.2).

The primary purpose of PRBSes is to provide clear cross-correlation between received and known transmitted signal. Besides this, selection of different PRB-Ses at each beacon inherently provides CDMA separation. Ie., the PRBSes can be transmitted and mixed together in the same, shared ultrasonic band. Separation improves with increasing sequence length (Eq. 2). On the other hand, longer sequence implies longer measurement period. Therefore, we decided to separate channels also by placing them into different frequency bands (FDMA).

2   Signal Flow

2.1   Modulation and Coding

Both transmitter and receiver in the system use Polaroid 600[1] electrostatic transducers. They offer relatively wide bandwidth ($\sim$ 30 kHz), compared to more common ceramic transducers (typically $\sim$ 2 kHz). The transducers require 150 V DC bias, and a transmitter should be excited nominally by 300 V_{p-p} AC voltage.

Frequency spectrum. Signals are transferred at ultrasonic frequencies 39 . . . 69 kHz, where the transducers offer the strongest response. BPSK (binary phase-shift keying) modulation is used for conversion of base-band PRBS signals into a ultrasonic band. BPSK of a square carrier[2] wave by a binary sequence produces

---

[1] Acquired and sold now by SensComp, Inc.

[2] Square carrier wave is a good replacement for ideal sinusoid in this case, as higher harmonics of the square wave lies far away from the signal band of interest.

also square wave as a result, simply generable by digital hardware, and suitable to be amplified to high voltage by a simple switching circuitry. Amplitude spectrum of a periodic PRBS of finite length is actually discrete, but in practice, it approaches a spectrum of infinite-length random binary sequence. After modulation by BPSK with carrier frequency f_0, neglecting side lobes mirrored about zero frequency, the resulting amplitude spectrum is $A(f) = A_0 \left| \operatorname{sinc} \frac{f-f_0}{f_b} \right|$, where $\operatorname{sinc} x = \frac{\sin \pi x}{\pi x}$, and f_B is a modulation frequency, ie. bitrate.

Although the main (effective) power content of the signal is transferred within a main lobe $f_0 \pm f_b$, there are still side lobes with amplitude $\leq \frac{f_b}{\pi(f-f_0)}$, interfering with a neighbouring channel. Besides the (slow) side-lobe amplitude decay, there is also an output filter in each transmitter, improving FDMA separation. To keep electronics simple, the output filter is only a 2nd order resonator (Sec. 2.2).

Available transducer bandwidth has been subdivided between signal bands (main BPSK lobes) and "safety gaps" between them. Chosen gap width resulted from trade-off between FDMA separation, and useful bandwidth, proportional to bitrate. The result is displayed at Fig. 1. $f_{1...3}$ are carrier frequencies of the three channels, corresponding to the three beacons. $f_b = 3\,\text{kHz}$, useful bandwidth per channel $\sim 2f_b = 6\,\text{kHz}$, distance between neighbouring carriers $f_2 - f_1 = f_3 - f_2 = 12\,\text{kHz}$.

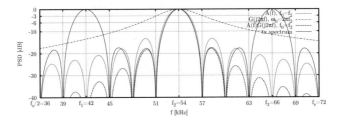

Fig. 1. Spectra of BPSK modulated PRBSes

All the carrier frequencies have been selected to be integral multiples of f_b: $f_1 = 42\,\text{kHz}$, $f_2 = 54\,\text{kHz}$, $f_3 = 66\,\text{kHz}$, and the carriers are coherent with transmitted PRBS. A sampling frequency $f_s = 72\,\text{kHz}$ in the receiver (Sec. 2.3) is also an integral multiple of f_b, although sampling is not coherent with transmission[3]. Also, all the three bands are placed symmetrically with respect to f_s, simplifying digital signal processing in the receiver (Sec. 2.3).

The transfer function of 2nd order filter is $G(s) = \frac{Bs}{s^2 + Bs + \omega_0^2}$, where $B\,[\text{rad}\,\text{s}^{-1}]$ is a 3-dB bandwidth. Given $f_0 = f_1$, $B = 2\pi 6\,\text{kHz}$, $\omega = 2\pi(f_0 + 12\,\text{kHz})$, an attenuation provided by the filter is $|G(s \approx j\omega)| \simeq 0.27 \sim 11\,\text{dB}$. Together with side-lobe decay, which is at 12 kHz distance $\approx \frac{1}{4\pi} \sim 22\,\text{dB}$, the FDMA channel separation is 33 dB.

---

[3] Because the receiver and transmitter clocks are not synchronized.

PRBS Codes. The PRBSes are transmitted continuously as a periodic signal, so the right tool to determine a time difference is a cyclic cross-correlation function

$$R_{xy}[n] = \sum_{m=0}^{N-1} x[m]y[(m+n) \bmod N] \tag{1}$$

of two signals $x[k], y[k]$[4] of period (length) N samples. An estimate of time displacement between x, y is $n_* = \arg\max_n R_{xy}[n]$.

PRBSes used in our system belong to set of Gold codes [3]. Numerical values ± 1 are assigned to PRBS bits. The auto-correlation function of a Gold code is $R_{pp}[0] = N, R_{pp}[n = 1 \ldots N-1] = -1$. Among other codes with the same property, the Gold codes have improved cross-correlation indifference between each other: different codes of the same length have their cross-correlation maxima, ie. false peaks in case of interference (crosstalk), bounded by some small value.

CDMA separation is given by a ratio of this cross-correlation upper bound and auto-correlation maximum. Gold codes are $N = 2^l - 1$ samples[5] long. According to [3], the ratio is

$$r_G = \frac{2^{\lceil \frac{l+1}{2} \rceil}}{2^l - 1}. \tag{2}$$

For $N \gg 1$ and l odd $r_G \approx \sqrt{\frac{2}{N}}$, for l even $r_G \approx \sqrt{\frac{4}{N}}$.

The only tunable parameter of Gold codes is the length N. It implies level of CDMA separation, and together with f_b also a measurement period T. Multiplied by a sound velocity c, T gives maximal measurement range D_{max}. We have chosen $N = 2^7 - 1 = 127$, what gives $T = 42.333$ ms, $D_{max} = 14.538$ m ($c = 343.4$ ms^{-1}). The period is much shorter than mechanical time constants of our robot, and the range also exceeds playground size.

For $N = 127$, the ratio of CDMA is $r_G = 0.13 \sim 18$ dB. Together with FDMA reserve, the overall channel separation is approx. 51 dB. Since radiated power is reciprocally proportional to squared distance, the minimal dynamic reserve given by minimal and maximal distances d_{min}, d_{max}, see Sec. 1.1, is $20 \log_{10} \frac{d_{max}}{d_{min}}$ dB $= 26$ dB. However, the "extra" dBs gained by FDMA & CDMA are often spent in fight against various imperfect signal propagation conditions, non-ideal processing in a receiver, curved transducer response, and lots of ubiquitous noise.

If a separation of 51 dB should be provided solely by CDMA, the corresponding Gold code length would be in order of 2^{17} or 2^{19}. More realistic choice may be a $N = 2^{11} - 1 = 2047$ code, providing 30 dB. Using such code and BPSK modulation over whole 30 kHz band, $f_b = 15$ kHz, would result in 3.22× longer (slower) measurement period. On the other hand, it would offer 5× higher baudrate, thus possibly a better distance measurement resolution.

In our system, we retained the combined FDMA & CDMA scheme, where the separation could be even increased by employment of higher-order transmitter output filters.

---

[4] If $y \equiv x$, R_{xx} is an auto-correlation function.

[5] Samples or bits, usually called "chips".

2.2 Transmitter

The beacons are interconnected by a wire, providing simple means of their synchronization. Square PRBS & BPSK signals for all three transmitters are generated by one ARM7 microcontroller using one integrated timer-counter.

Microcontroller is set at one beacon, distributing square-wave signals to two other beacons using physical layer of RS485 interface standard. Logic-level square signals are current amplified at 5 V supply by a buffer and transformed to $\sim$ 200...300 V_{p-p} by a custom pot-core transformer. The transformer coil is a base of the 2nd order LC filter, formed together with a transducer, bandwidth defining resistor, and a tuning capacitor. Another transformer and rectifier provides a bias voltage.

2.3 Receiver

Sampling. Signal flow from the transmitter to the receiver and an estimator is shown at Fig. 2. The three signals travel through the air, and delayed by the path length, they are mixed and received by an ultrasonic transducer at the robot. Received signal is directly sampled by an analog-to-digital converter (ADC) out of the baseband, ie. at sampling frequency f_s lower than double the maximal input frequency. Thanks to the out of the baseband sampling, no front-end downconversion is necessary. On the other hand, a band-pass anti-aliasing filter is necessary instead of common low-pass one. The band-pass filter has been realized as an active 8th order Chebyshev type I, 1 dB ripple filter, built using operational amplifiers.

Input band 39...69 kHz sampled at $f_s = 72$ kHz is aliased (mirrored) as 33...3 kHz. High voltage DC bias of the receiving transducer is produced by a transformer, switched at f_s. The switching frequency and its harmonics are aliased at either zero frequency or $\frac{f_s}{2}$, where no useful signal is present.

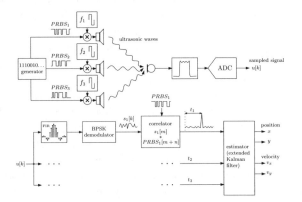

Fig. 2. System diagram and signal flow from beacons to receiver

The sampling and further signal processing at receiver side is by no means synchronized with the transmitter – there are two independent clocks, each driven by its own crystal oscillator.

Filtering. The sampled digital signal is then subdivided into three 6 kHz wide bands, centered around aliased carrier frequencies, $f_1' = 30\,\mathrm{kHz}, f_2' = 18\,\mathrm{kHz}, f_3' = 6\,\mathrm{kHz}$. A 48th order FIR filters have been designed using least-squares approximation of amplitude frequency response, including transition bands, symmetrically for all the three filters.

Filter coefficients have been quantized into $-127 \ldots 127$ range to allow efficient hardware or software implementation by an $8 \times n$-bit multiplier[6]. Thanks to band symmetries, for each two non-zero filter coefficients at k-th position, equation $b_1[k] = \pm 2^m b_2[k]$ holds for some $m \in \mathbb{Z}$. Moreover, for any linear-phase FIR filter, an even or odd symetry between coefficients holds.

As a result, only 19 multiplications and 73 additions or subtractions, accompanied by bit shifts, suffice to filter one sample by all three FIRs. On ARM7 CPU, the whole filtering period including memory transfers and wait states takes 350 CPU cycles. This gives a 60 MHz microcontroller 42% load at $f_s = 72\,\mathrm{kHz}$.

BPSK Demodulation. Each filtered signal is then BPSK demodulated. Since f_s is integral multiple of f_b, and delays are easy to implement in discrete-time domain, a differential BPSK demodulator has been used. Its (nonlinear) difference equation is $y[k] = x[k]x[k - M]$, where k is a time index, and $M = \frac{f_s}{f_b}$ (number of samples per PRBS bit). Demodulator should be followed by a low-pass filter to suppres upper frequency modulation products. A simple moving-average filter, ie. summing last M samples, has been used.

However, the differential demodulator does not output the same sequence as the modulated one. In effect, each two successive original bits were XOR-ed to produce an output bit: $p[n] = q[n - 1] \oplus q[n]$ ($q[n]$ being the original, modulated sequence, $p[n]$ the output, demodulated sequence, and n is the bit number modulo sequence length). Therefore, transmitted sequences have been altered to produce the Gold code PRBSes after differential demodulation.

Demodulated and roughly filtered signal is then decimated, because the effective bandwidth of a binary sequence in baseband is $1.5 \ldots 3\,\mathrm{kHz}$. New sampling rate $f_{s_2} = \frac{f_s}{6}$, ie. 4 samples per bit[7].

Correlation. Demodulated and decimated signals enter the correlation process. After collecting one PRBS period, ie. $N = 4 \times 127$ samples (at this point, bits are oversampled), a peak in correlation between received and known transmitted sequence should reveal the time displacement.

---

[6] Eg. ARM7 CPU core contains a multiplier, which performs 8×32 bit multiplication in one cycle; the another advantage is, that 8-bit value fits into an immediate machine-code instrucion operand value, thus a memory access is saved.

[7] In theory, 2 samples per bit should be sufficient; faster rate has to compensate poor filtering properties of the moving-average filter.

Before correlation, the whole period of N samples is buffered and normalized to 8-bit range (software "automatic-gain-control"). Then, the correlation $R_{xy}[n]$ should be evaluated for all possible displacements $n = 0 \dots N - 1$.

Computation of correlation for all $n = 0 \dots N - 1$ following definition (1) involves N^2 multiplications of x and y and $\sim N^2$ additions. Since one of the signals is composed of ± 1, the multiplication reduces to conditional sign change of the second (integer) number. The correlator performs N^2 conditional sign changes and additions. In our application, $N = 4 \times (2^7 - 1) = 508$, so the correlator performs 258064 such operations in one period.

Of course, there exist much more efficient methods to compute the correlation, but they are defined only for sequences with too restrictive properties.

If both $x, y \in \{\pm 1\}$, the multiplication reduces to bitwise boolean XOR. Complexity remains $\mathcal{O}(N^2)$, but implementable in very fast and vectorized fashion on all architectures. However, due to large dynamic range of signals in our system (26 dB), the 1-bit instead of 8-bit quantisation can lead to unusable results.

The correlation can be also very efficiently computed using fast Fourier transform (FFT) algorithm in $\mathcal{O}(N \log N)$. However, the FFT is not applicable for prime-length sequences and almost useless for sequences of length, factorized into few primes. This is exactly the case of all Gold codes. In our case, the signal length is prime-factorized as $508 = 2 \times 2 \times 127$. We could change the sequence length to eg. $4 \times 2^7 = 512$, but the correlation properties of PRBSes would degrade.

Software implementation of the $\mathcal{O}(N^2)$ correlator took ~ 20 ms of CPU time on 400 MHz PowerPC 603e. Given the 42.333 ms PRBS period, the real-time CPU load is 47%. As an alternative, we have developed custom FPGA correlator core as a dedicated computational resource. The correlator has been implemented into Xilinx Virtex-4 XC4VFX12 FPGA. Running at 300 MHz clock, the correlation for $N = 508$ took 0.215 ms. Compared to PowerPC, the processing is done by the correlator FPGA core 93× faster. Moreover, the correlation is performed in parallel, so it does not spend any CPU time.

Interpolation. The correlator output $R_{xy}[n]$ should contain an accentuated peak above a noise floor. A position n_* of the peak is the time displacement, quantized to $\frac{1}{f_{s2}}$. To obtain finer resolution, $R_{xy}[n]$ samples are interpolated.

It follows from the Wiener-Khinchin theorem, that iff one of signals x, y is band-limited, then their correlation R_{xy} is band-limited as well. The received signal is band-limited, although in a non-ideal way. According to the sampling theorem, a continuous correlation function $R_{xy}(t)$ can be fully reconstructed from the sampled one, $R_{xy}[n]$, by ideal low-pass interpolation.

Position of $R_{xy}(t)$ maximum can be an arbitrary real value, t_*. Actually, the precision of t_* is limited by noise in $R_{xy}[n]$. Therefore, the ideal interpolation is fruitless[8], an approximate FIR low-pass interpolation using few samples around n_* is a relevant method.

---

[8] Although it is possible to find maximum of ideally interpolated $R_{xy}[n]$, eg. using Newton method.

Interpolating FIR filter has been designed using windowing method, spanning 5 neighbouring samples. Interpolating ratio $32 : 1$ gives resolution $\frac{D_{max}}{32N} = 0.9\,\text{mm}$, far better than precision of mechanical system parts, so there is no need of finer interpolation.

Each of the three interpolators produces the interpolated correlation peak position, t_*, denoted $t_{1...3}$, at a sampling period $T = 42.333\,\text{ms}$. These values enter a position estimation (or calculation) process.

3   Position Estimation

3.1   Calculation vs. Estimation

The final task of the localization system is to convert measured times $t_{1...3}$ to robot coordinates, $x_{1...2}$. The times $t_{1...3}$ are equal to distances[9] between robot and respective beacons, plus a transmitter-receiver clock offset, Δ:

$$t_i = \sqrt{(x_1 - x_{Bi})^2 + (x_2 - y_{Bi})^2} + \Delta, \tag{3}$$

where x_{Bi}, y_{Bi} are coordinates of i-th beacon. As mentioned in Sec. 2.3, both clocks are running independently on their own crystals, so they are in each measurement biased by some fraction of period, $\Delta \in [0, D_{max})$. The task can be solved by one of the following ways.

Direct Calculation. The system of three equations (3), $i = 1 \ldots 3$, is regular, containing three unknown variables, $x_{1...2}$ and Δ. First, equations are subtracted to eliminate Δ – only a time differences, $t_3 - t_1, t_2 - t_1$, remain. The problem is called TDOA (time differences of arrivals) or hyperbolic navigation, since in geometric insight, it relies in intersection of two hyperbolas [4, 5].

The system yields a quadratic equation. One of the two solutions is incorrect. It may be often eliminated after substitution into original equations (3). Otherwise, if both of the solutions are feasible, the choice depends on an external knowledge about the robot.

Under an assumption of absolutely precise measurement of $t_{1...3}$, the result is exact. However, it never holds in practice, since the measurements suffer from uncertainity (noise). This is the reason, why the stochastic methods give much more precise results, than the direct calculation.

In case of noisy measurement, the quadratic equation may have no real solution. It can be overriden by forcing its discriminant to zero.

Although the method does not provide the best possible precision, it can provide an initial guess for the following methods.

Static Stochastic Estimation. Another possibility, how to infer $x_{1...2}$ from $t_{1...3}$, is to minimize (numerically) some norm of the uncertainity in $x_{1...2}$ estimate, eg. $\sigma_{x_1}^2 + \sigma_{x_2}^2$, according to the error propagation law, given t_i variance, σ_t^2.

---

[9] In the following, we assume constant sound velocity c, therefore expressing all time quantities in units of length.

Such an optimization should give slightly better results, than the direct calculation. However, it still does not exploit a substantial property of the system: its dynamics.

Dynamic Stochastic Estimation. The dynamic estimator is supplied with a system model – besides the static information about measurement uncertainity (noise), the model contains information about evolution of variables (state) in time, including uncertainity of the evolution (process noise).

The static stochastic estimation can be regarded as a weighted sum of several contributions with different variances. The dynamic estimation then extends this idea: it weights not only currently measured values, but also past values. Due to the process noise, older values have increased covariance, and therefore lower weights than the recent values.

Roughly speaking, the state variables correspond to a memory (inertia) of the system. In our system, there are two distinct dynamic subsystems: robot motion dynamics, and clock dynamics (Sec. 3.2). The essential improvement in precision over the static estimation consists in exploiting the latter, since the process noise of crystal oscillators, producing Δ, is very low.

Systems that estimate Δ and then determine position by calculation or static estimation are known as TOA (times of arrivals), or spherical navigation systems [4, 5]. Such a case is analogous to oscillator phase-locked loop (PLL), however, the "phase" (Δ) differences are not evaluated explicitly, but they are contained in coumpound measurement, $t_{1...3}$. However, the estimator may profit from motion dynamic model as well.

If a good estimate of Δ is known, the system of three equations (3) is overdetermined in effect. Therefore, the position can be still calculated, if one of the $t_{1...3}$ measurements is missing. This is very important, because one of the beacons can be always occluded by a competing robot. The system should survive "locked" to Δ and its drift for several seconds, while one of the three measurements is missing.

The estimator can evaluate likelihood of individual measurements, thus it may provide simple means of outlier detection, caused eg. by beacon occlusion, or reception of a stronger reflected signal.

The estimator can also process data from other external sources (sensors and actuators), to perform a data-fusion. It can either improve precision, widen bandwidth, or provide additional estimated variables, such as a heading φ by cooperation with odometry.

3.2 System Model

Stochastic System. Design of dynamic estimator consists in choice of system model, and an algorithm. General model of nonlinear stochastic system is

$$x[k + 1] = f(x[k], u[k], v[k]) \qquad (4)$$

$$y[k] = g(x[k], u[k], e[k]) \qquad (5)$$

where k is a discrete time, x the system state, u a known input signal (where applicable, eg. motor actuation), v the process noise, e the measurement noise, and y is a measurement (all values real and possibly vector). Difference equation (4) describes system dynamics, and g is an output function. The noises are specified by their stochastic properties, eg. as uncorrelated Gaussian white noises with zero mean and covariances $\mathcal{E}\{vv^T\} = Q, \mathcal{E}\{ee^T\} = R$.

In our design, we managed with a much less general and simpler model

$$x[k + 1] = Ax[k] + v[k] \tag{6}$$

$$y[k] = c(x[k]) + e[k] \tag{7}$$

– the only nonlinearity is $c(x[k])$, based on (3). The dynamics is linear, as well as contribution of both noises. No known system inputs are considered, all information is propagated through the measurement.

Integrator Model. An integrator is a simplistic model of 1st order system with infinite time constant. It expects, that $x[k]$ will be near previous $x[k-1]$, with uncertainity given by the process noise. In terms of (6), $x = (x), A = (1)$. If x is expected to move with almost constant velocity (or drift) v, it can be modelled by double integrator: $x[k] = x[k-1] + v[k-1], v[k] = v[k-1]$, or $x = (x, v)^T, A = \left(\begin{smallmatrix} 1 & 1 \\ 0 & 1 \end{smallmatrix}\right)$.

The single or double integrator submodels have been used in our system for both clock and motion dynamics.

Clock Dynamics. Clock offset Δ modelled by single integrator expects, that clock drift $\Delta' = \frac{\partial \Delta}{\partial t}$ is small. Better is to use the double integrator, estimating Δ' along with Δ. Drift is then considered to be constant, varied by an unmeasured process noise (mostly temperature changes in reality).

Motion Dynamics. Precise motion dynamic model contains mechanical time constants given by mass, inertia, and frictio. Moreover, the motion dynamics often contains substantial kinematic nonlinearities (anisotropic friction of wheels, dependent on heading φ). The motor actuation signal may, or may not be included in the model. Obtaining model parameters may be difficult (system identification task).

On the other hand, the integrator dynamics offer simple, although suboptimal approach. It can not gain as good precision as a more realistic model. However, it describes well the situations, where acceleration is occasional and robot remains in unchanged motion most of the time. The double integrator motion models are popular, where a very little is known about object dynamics (computer vision, tracking, [6]).

Measurement Model. Ultrasonic measurement is modelled by the function $c(x)$ with an additive noise e. The noise represents uncertainity of correlator and interpolator output. Its variance is fixed.

Odometry Fusion. Cartesian coordinates $x_{1...2}$ may be estimated from ultrasonic readout $t_{1...3}$. To obtain an estimate of heading φ, we have added odometry, ie.

data from wheel incremental sensors, to the system. Such a data fusion of relative positioning (odometry) with an absolute one (ultrasonic system) is a common procedure in mobile robotics [7–9].

Essentially, the increments over fixed period of time represent wheel velocities, v_L, v_R. Obviously, a trajectory reconstructed from these velocities suffer from cumulative error, and it is up to the estimator to correct it using absolute measurements.

Very often the odometry is treated as a known actuation signal u to the system, and the absolute positioning information is treated as a measurement y, [7–9]. However, the local linearization (Sec. 3.3) of such a model with additive process noise is not observable – φ remains obscured by cumulative noise of odometry. Neither the transition matrix A, nor an output $c(x)$ provide coupling between φ and a rest of the state vector, therefore it can not be corrected by ultrasonic measurements.

To overcome this problem, we treat the odometry as a second measurement $y_2 = (v_L, v_R)^T$, beside an ultrasonic measurement $y_1 = (t_1, t_2, t_3)^T$, $y = (y_1^T y_2^T)^T$. The system is considered to be actuated only by the process noise. This approach seems to be more realistic, as the odometry is treated as a velocity measurement with additive noise. Using this model, all the system states are observable, including φ and its derivative, angular velocity ω.

The final 8th order model, based on $4\times$ double integrator dynamics, see Fig. 3, is:

$$x = (x_1, v_1, x_2, v_2, \varphi, \omega, \Delta, \Delta')^T$$

$$Q = \text{diag}(0, \sigma_x^2, 0, \sigma_x^2, 0, \sigma_\omega^2, 0, \sigma_\Delta^2)$$

$$A = \begin{pmatrix} 1 & 1 & 0 & 0 & 0 & 0 & 0 & 0 \\ 0 & 1 & 0 & 0 & 0 & 0 & 0 & 0 \\ 0 & 0 & 1 & 1 & 0 & 0 & 0 & 0 \\ 0 & 0 & 0 & 1 & 0 & 0 & 0 & 0 \\ 0 & 0 & 0 & 0 & 1 & 1 & 0 & 0 \\ 0 & 0 & 0 & 0 & 0 & 1 & 0 & 0 \\ 0 & 0 & 0 & 0 & 0 & 0 & 1 & 1 \\ 0 & 0 & 0 & 0 & 0 & 0 & 0 & 1 \end{pmatrix} \quad (8)$$

$$y = (t_1, t_2, t_3, v_L, v_R)^T$$

$$R = \text{diag}(\sigma_t^2, \sigma_t^2, \sigma_t^2, \sigma_{LR}^2, \sigma_{LR}^2)$$

$$c(x_1, x_2, \Delta, v_1, v_2, \varphi, \omega) = (c_{11}, c_{12}, c_{13}, c_{2L}, c_{2R})^T \quad (9)$$

$$c_{1i} = \sqrt{(x_1 - x_{Bi})^2 + (x_2 - y_{Bi})^2} + \Delta \quad (10)$$

$$c_{2L,R} = \frac{1}{2}(v_1 \cos \varphi + v_2 \sin \varphi \pm k_\omega \omega) \quad (11)$$

3.3   Estimation Algorithm

Nonlinear Estimation Methods. Estimation of a state of a linear system, ie. a system defined by (6, 7) with linear output $c(x) = Cx$, is a straightforward task, solved by linear filtering. The linear estimator, performing filtering of data $u[k], y[k]$ and producing $x[k]$, is called observer in general, and the Kalman filter, if optimized with respect to Q, R covariances. In case of nonlinear system, there are several approximate estimation methods, from which we have tried two: the particle filter, and the extended Kalman filter (EKF).

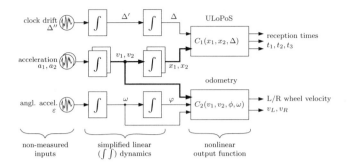

Fig. 3. Block diagram of $4\times$ double integrator system model with both ultrasonic and odometric measurement

In both of the estimators, state is represented by a probability density function (PDF) $P(x)$. Measurement is represented by a likelihood function $L(x|y) = P(y|x)$ (function of x for known y). Within each sampling period, $P(x)$ evolves in two steps. In prediction step, $P(x)$ is transformed and convolved with v according to (4). After new measurement, corrective step arises. $P(x)$ is multiplied with $P(y|x)$ and normalized, where $P(y|x)$ is given by (5).

Particle Filter. Also known as Monte Carlo estimator [10], approximates $P(x)$ by a large set of randomized points (particles) in state space. In the prediction step, each particle is transformed by (4) using randomly generated v. In the correction step, particles are weighted by $P(y|x)$, then replicated and eliminated to form next $P(x)$.

Particle filter provides global PDF approximation. Therefore, it does not need intial guess in situations with large prior uncertainity (startup, error states). Also, it allows to preserve any non-Gaussian, even multimodal PDFs, what may slightly help during erroneous measurements.

The major disadvantage of the particle filter is its computational exigency. While processing real data by 4th order estimator, the number of particles required for smooth filtering was $\sim 10^4$. It has been impossible to reach real-time estimation using C language implementation on embedded computers available in our project (400 MHz PowerPC 603e). Although the estimation ran only offline, it has been useful during system design. The particle filter has been implemented according to [10] (Sec. 4.1, Algorithm 1), applied to models with 1st order motion dynamics together with 1st or 2nd order clock dynamics.

Extended Kalman Filter. EKF [11], is based on local linearization. $P(x)$ is considered to be Gaussian, desribed by mean x and covariance matrix P. Equations (4, 5) are linearized in point $x = x[k]$, $u = u[k]$, $v = e = 0$. The resulting linear system is used to estimate next $x[k+1]$, $P[k+1]$, like in an ordinary linear Kalman filter. Gaussian variances are assumed to be small enough with respect to local linearization.

Since the approximation is only local, it requires a feasible initial state estimate to work properly. Otherwise, convergence is not guaranteed.

EKF can not cope with arbitrary PDFs. This does not really matter in our system, as during proper operation, all measurement PDFs intersect together in one narrow peak. An ambiguous multimodal PDF can occur, when some measurements are missing (because considered faulty), or when a fusion with ambiguous sensory data (eg. local playground color) would be requested.

To eliminate both of the drawbacks mentioned, the estimator is equipped with error recovery procedure (see below).

Algorithm of EKF used in frame of our system follows. The local linearization of the model (6, 7) consists in linearization of output function (9):

$$y[k] \approx c(\hat{x}) + C(\hat{x})(x - \hat{x}) + e\,, \qquad C(\hat{x}) = \left.\frac{\partial c(x)}{\partial x}\right|_{x=\hat{x}} \qquad (12)$$

where $\hat{x}$ is a current state estimate (linearization point), and $C(\hat{x})$ is linearized output matrix. The prediction step follows linear dynamics:

$$x'[k] = Ax[k-1]\,, \qquad\qquad P'[k] = AP[k-1]A^T + Q \qquad (13)$$

Then, the correction step using $C(x'[k])$ is performed:

$$C[k] = C(x'[k])\,, \qquad\qquad M[k] = C[k]P'[k]C^T[k] + R \quad (14)$$

$$L[k] = P'[k]C^T[k]M^{-1}[k] \qquad\qquad (15)$$

$$x[k] = x'[k] + L[k](y[k] - c(x'[k]))\,, \quad P[k] = P'[k] - L[k]M[k]L^T[k] \quad (16)$$

First, $P(x)$ mean x is shifted by the transition matrix, then, covariance P is transformed according to the error propagation law, and widened by the process noise in (13). The optimal Kalman gain matrix L is calculated in (15). A difference between measured output y and modelled output $c(x')$ is multiplied by the Kalman gain, and added as a correction to x in (16), and P is narrowed by information contribution of the measurement.

EKF has been implemented in C language, floating-point math. Inversion involved in (15) is calculated using Cholesky factorization, since $M = M^T, M > 0$. There are two operations, vulnerable to irregular input data: square-root, and division. They have been protected against negative values, and near-zero divisor, respectively.

Application Specific Details. EKF must be provided with feasible initial state. Therefore, the initial state is set to $x[0] = (x_1, 0, x_2, 0, 0, 0, \Delta, 0)^T$, where x_1, x_2, Δ are the closed-form solution of TDOA (Sec. 3.1).

The ultrasonic measurement subsystem does not provide self-reliant detection of erroneous measurements. Actually, it is almost impossible to distinguish between correct and false measurement, eg. in case of direct signal path occlusion and reception of a reflected beam. Therefore, the error detection relies on the estimator. Obviously, it is not able to reveal all possible errors, because some of the erroneous values may lie close enough to correct ones, and confuse the detector.

Likelihood of individual i-th measurement $P(y_i|\boldsymbol{x})$ can be evaluated before the correction step. A similar, yet more simply computable measure, excentricity ε, has been used to judge faulty measurements. In the correction step (16), contribution of i-th measurement y_i to the state is $\Delta\boldsymbol{x} = \Delta y_i \boldsymbol{l}_i$, where Δy_i is difference between measured and predicted y_i, and $\boldsymbol{l}_i$ is i-th column vector of $\boldsymbol{L}[k]$. ε is a ratio of $|\Delta\boldsymbol{x}|$ to standard deviation of $\boldsymbol{x}$ in direction of $\Delta\boldsymbol{x}$:

$$\varepsilon^2 = \Delta y_i^2 \boldsymbol{l}_i^T \boldsymbol{P}^{-1}[k] \boldsymbol{l}_i \tag{17}$$

If ε is greater than specified threshold ($\varepsilon^2 > \varepsilon_{thr}^2$, for eg. $\varepsilon_{thr} = 30$), the measurement is considered faulty (too unlikely). Corresponding column $\boldsymbol{l}_i$ is reset to zero, to protect state vector against damage.

While only one of the three ultrasonic measurements is considered faulty, the system should continue operation. Position estimation should manage with estimated clock offset and drift. It may be advisable to set temporarily the clock process noise σ_Δ^2 to zero.

If two or more measurements are evaluated as faulty, the system may get lost (diverge) due to accumulated errors. When the system detects more than N_{lost} successive iterations with more than one faulty measurements, it declares being lost, and tries to reinitialize next step with a new initial state.

4 Results

System Performance. Errors occured, when ultrasonic signal reflections were stronger than signal received by direct path. If the reflections were weaker, then they made no problem, since corresponding correlation peaks are lower. We have not been able to measure and adjust directional characteristics of the transducers to eliminate the problem entirely. In our experimental room, there sometimes occured stronger reflections from plastic files on a shelf, too near to playground boundary. A woolen sweater cast onto the files solved the problem.

In some areas, occasional noisy measurements occured, causing temporary deviations of measured position of $\sim 1 \ldots 10$ cm. It may have been incurred by slightly broken beacon mechanics. But most of the time, the measured position or trajectory were apparently clean.

Precision during robot movement has not been evaluated yet, due to lack of a suitable reference positioning system.

Position measured at stationary point shown standard deviations $\hat{\sigma}_{x_1} = 1.8$ mm, $\hat{\sigma}_{x_2} = 4.4$ mm. Compared to manually measured position, maximal error was 17.6 mm and mean bias 4.5 mm during 84 samples long measurement. It should be pointed out, that mounting precision of mechanical parts (beacon holders) has been $\sim \pm 10 \ldots 25$ mm. Real conditions at a competition may be similar.

Conclusion. In stationary measurements, the precision is in order of mounting precision of playground parts. The immediate following step is to close a positional feedback to robot motion control along a specified trajectory.

Next, the system precision should be evaluated using camera system, and maybe also a pencil-paper trajectory recording.

In the future, we may try to search another type of transducers or reflectors, to provide better and easier beam alignment.

Work supported by the Ministry of Education of the Czech Republic, project 1M0567.

References

1. Eurobot: EUROBOT 2009: Temples of Atlantis (2009), http://www.eurobot.org/
2. Chudoba, J., Přeučil, L.: Localization using ultrasonic beacons. In: 3rd International Congress on Mechatronics 2004, Praha, ČVUT v Praze, FEL (2004)
3. Linnartz, J.P.M.G.: Wireless Communication. Kluwer Academic Publishers, Dordrecht,
   http://www.wireless.per.nl/reference/chaptr05/cdma/codes/gold.htm
4. Linde, H.: On Aspects of Indoor Localization. PhD thesis, Fakultät für Elektro- und Informationstechnik, Universität Dortmund (August 2006)
5. Jiménez, A.R., Seco, F., Ceres, R., Calderón, L.: Absolute localization using active beacons: A survey and IAI-CSIC contributions. White paper, Instituto de Automática Industrial – CSIC, Madrid, Spain (2004)
6. Davison, A., Reid, I., Molton, N., Stasse, O.: MonoSLAM: Real-Time Single Camera SLAM. IEEE Transactions on Pattern Analysis and Machine Intelligence 29(6), 1052–1067 (2007)
7. Fox, D., Burgard, W., Dellaert, F., Thrun, S.: Monte Carlo localization: Efficient position estimation for mobile robots. In: Proc. of the National Conference on Artificial Intelligence (1999)
8. Winkler, Z.: Barbora: Creating mobile robotic platform. In: MIS 2003. Matfyzpress (2003)
9. Köse, H., Akın, H.L.: The reverse Monte Carlo localization algorithm. Robotics and Autonomous Systems 55(6), 480–489 (2007)
10. Carpenter, J., Clifford, P., Fearnhead, P.: Improved particle filter for nonlinear problems. IEE Proceedings Radar, Sonar and Navigation 146(1), 2–7 (1999)
11. Ljung, L.: Asymptotic behavior of the extended Kalman filter as a parameter estimator for linear systems. IEEE Transactions on Automatic Control 24(1), 36–50 (1979)

Calibration Methods for a 3D Triangulation Based Camera

Ulrike Schulz[1] and Kay Böhnke[2]

[1] Baden-Wuerttemberg Cooperative State University Mannheim, Germany
[2] University of Applied Science, Heidelberg, Germany

Abstract. A sensor in a camera takes a gray level image (1536 x 512 pixels), which is reflected by a reference body. The reference body is illuminated by a linear laser line. This gray level image can be used for a 3D calibration. The following paper describes how a calibration program calculates the calibration factors. The calibration factors serve to determine the size of an unknown reference body.

Keywords: calibration, 3-D camera, laser, triangulation.

1 Introduction

The following experiments and solutions in this paper deal with a 3D laser measurement system employing the high speed camera "Sick Ranger E55" and the associated laser diode. A detailed technical description concerning the camera can be found in [1]. The Ranger camera creates a measurement of a reference body, which passes during the measurement process the field of view of the camera with constant speed. The reference body is illuminated by the laser. While measuring is carried out, the sensor in the camera collects the reflected light and the dispersion of the reference body, which can be analyzed in the following. The used laser is a low power device, whose output does not exceed 1mW and it therefore belongs to Class 2/2M. The wavelength of the emitted light is in the visible spectrum. For the 3D measurement a conveyor belt is necessary, which moves the reference body at constant speed below the camera. Thus it is possible to capture the body's Y-coordinates with the camera. The height is derived from the Z-coordinates and the width measured by use of the X-coordinates. The results of the measurement can be used to determine a calibration factor of the different coordinates. With the determined calibration factor the measurement of unknown bodies is possible. Before a measurement can be made, the following criteria must be fulfilled:

1.1 Correct Camera and Laser Configuration

For the arrangement of the camera and the laser, there are four different possibilities [2]:

1. Reversed Ordinary: This means, the laser is positioned orthogonally above the reference body. The camera is mounted at a defined angle to the laser. It can be stated, that this arrangement provides a good height resolution.

A. Gottscheber, D. Obdržálek, and C. Schmidt (Eds.): EUROBOT 2009, CCIS 82, pp. 131–143, 2010.

2. Ordinary: In this case, the camera is mounted orthogonally above the reference body and the laser is arranged at an angle to the camera. This setup provides the maximum resolution. In this configuration, the laser line is projected horizontally on the reference body, so that the measured values can be assigned to the different lines taken by the camera.

3. Specular: A specular setup means that laser and camera are opposite to each other and both at an angle to the reference body. The height resolution achieved with this configuration is not useful, because the reference body is covered in shadows at its front and its end. Nonetheless it is good for surface analysis [3].

4. Look away: In this setup, laser and camera are on the same side of the reference body, but both at an angle to it. The reflection of the laser light returned by the reference body is not good enough for a sufficient height resolution, as the laser light is too low-powered [3].

The result is that the displacement of the laser depends on the geometric position of the camera. If the camera is mounted at a high angle to the reference body, the resolution and the shift of the laser line are better than in case of a low angle. An inappropriate configuration for the laser and camera is given, if the laser and camera are positioned at the same side and at the same angle to the reference body. For this arrangement a more powerful laser is needed. The following pictures show the different possibilities of assembling with the associated formula for determining the height:

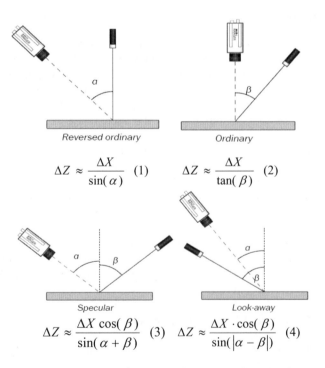

$$\Delta Z \approx \frac{\Delta X}{\sin(\alpha)} \quad (1) \qquad \Delta Z \approx \frac{\Delta X}{\tan(\beta)} \quad (2)$$

$$\Delta Z \approx \frac{\Delta X \cos(\beta)}{\sin(\alpha + \beta)} \quad (3) \qquad \Delta Z \approx \frac{\Delta X \cdot \cos(\beta)}{\sin(|\alpha - \beta|)} \quad (4)$$

Fig. 1. Different camera-laser setups ("Reversed Ordinary"; "Ordinary"; "Specular" and "Look away") and the associated formulas for determining the height [2]

1.2   Correct Configuration of the Ranger Camera

Necessary components, functions and parameters for the configuration of the camera are:

The range of the sensor ("Region of Interest (ROI)"), which is in the focus for the measurement. The SICK Ranger E55 camera can use a maximum field of 1536 columns for the ROI width and 512 rows for the ROI height. For simultaneous measurements different regions of interest can be defined, which are not allowed to overlap each other, though [4].

Additional light sources are needed to ensure an accurate measurement. The sensor evaluates the center of the reflected laser line searching the row with the highest light intensity in every column and thus determines the position of the reference body.

The choice of the measurement method affects the measurement of height and should be chosen in such a way, that height, width and length can be measured accurately. Additionally, the length measurement depends on the form of the trigger (e.g., if the height profile is delivered at the beginning or the end of the pulse).

A problem that occurred during the calibration (width, height and length measurement) is the fact, that the sensor is only capable of measuring a number of pixels. To process the captured data, the pixel-number needs to be converted in millimeters. The arrangement of camera and laser is important for this procedure, because an accurate measurement should be optimally covering the pixel area. I.e. if the distance to the reference body is small, the height, width and length can be calibrated more accurately (better resolution). Another item is the brightness of the camera, which can be influenced by the exposure time and aperture number. The exposure time returns the span of recording time for the camera. The longer the exposure time, the brighter and more intense will be the shade of gray used for measurement. Short exposure times can be realized with low aperture numbers. An incorrect setting of these two components can lead to over- or under-exposed and blurred images.

2   The Idea of Calibration

2.1   Choice between Image Profile and Measurement Profile

There are two different ways of recording the reference body with the camera: image mode and measurement mode. With the image mode only 2D measurements are possible, because the image mode only records 2D images. In everyday life the image profile is used for image viewing applications. The measurement profile additionally returns the altitude profile of the laser line. The measurement mode is not usable to measure a body if the camera and the laser are mounted at the same position (see above). The measurement mode is only issued on one line and that is the line where the laser is located. It is possible to detect height and width from only one measurement profile. The height is indicated by the shade of gray, e.g. a dark gray (close to black) matches a low gray value, vice versa, a bright gray matches a high gray value. This value serves as an indicator for the distance between the reference body and the laser. If the reference body is close to the laser then the reflection is more intense and the shade of gray is bright. The width can be calculated as the distance between two identical shades of gray, complying to the conveyor belt. The measurement profile can be scanned profile after profile from the moving reference body. The addition of all height profiles produces a 3D model of the reference body.

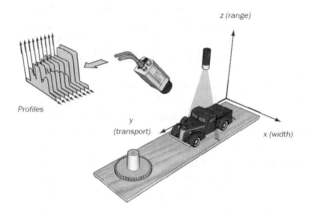

Fig. 2. Measurement process in measurement mode and addition of the height profiles to obtain the 3D-model [5]

Within the measurement mode various methods exist for the height measurement/calibration of the reference body, which can be used for further analysis.

5.   Threshold: Every pixel, whose value exceeds the threshold, is set to 1, every pixel whose value is below the threshold, is set to 0. A possibility to set the threshold is to use the arithmetic mean of the highest and the lowest gray value.

6.   Maximum: This means the representation of the row where the highest intensity of light has occurred. I.e. only the highest point is valid.

7.   Center of gravity: This method uses a threshold as well. The focus lies on the area above the threshold and the main emphasis defined. In (5), the row in which the sensor is located is called X and $f(X)$ is the intensification of light which is received by the sensor [6].

$$\text{Center of gravity} = \frac{\sum (X \cdot f(X))}{\sum f(X)} \tag{5}$$

The best height measurement method is "Hi3D" with the usage of center of gravity. This method uses a special algorithm for valuation of the emphasis of the intensity. Thus a result with a resolution of 1/16 subpixel and high accuracy can be achieved [7].

2.2 Measurement of a Reference Body

The evaluation of the reference body takes place in three steps:

1.   First of all the width has to be measured. Therefore, the edges of the reference body have to be identified. The upper and lower edges are indicated by pixel positions (columns). The difference between the two edges is the searched width.

2.   In the second step the height is measured. The Height is determined from the displacement of the laser line in the measurement mode. The measurement mode collects a shade of gray from each pixel of the laser line. The shade of gray is a measure for the height. Prerequisite for the height measurement is that the position of the reference body has already been determined. The height measurement takes place in the middle between upper and lower edge. At this fixed position the shade of gray is detected with and without reference body. The difference between these two shades of gray determines the height of the reference body.

3.   By the time width and height are determined a frame counter can capture the length of the reference body. By using the width and height information, a mask is created and compared with the incoming images. For this part of the process, the reference body is located on the moving conveyor belt and thus passes below the camera. If the received image fits the mask, the frame counter increments by one. During the start of the program the frame counter is set to zero. The total number of fitting profiles determines the length of the body.

2.3 Implementation of the Calibration

Three calibration factors are determined by the calibration process. These calibration factors make it possible to measure an unknown body and provide the output value in SI units. As mentioned above, the accuracy of the calibration depends on the position of the camera and laser. This can be mathematically deduced from the laser triangulation. The picture shows the "Reversed Ordinary" configuration with the appropriate designations for the derivation of the formula for height measurement.

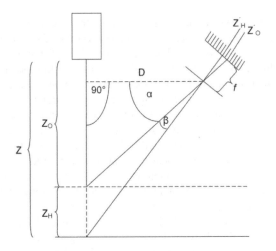

Fig. 3. Using the arrangement "Reversed Ordinary" and showing all the distances and angles one needs to know for the laser triangulation

The requested height is labeled as Z_H, which is one part of the segment Z. First of all the formula for Z is set up.

$$Z = D \cdot \tan\left(\arctan \frac{Z_O}{D} + \arctan \frac{Z' - Z'_o}{f} \right) \tag{6}$$

According to the addition theorems results
  Thus solving the equitation for Z_H, arises the final formula:

$$\tan(\alpha \pm \beta) = \frac{\tan \alpha \pm \tan \beta}{1 \mp \tan \alpha \cdot \tan \beta}. \tag{7}$$

$$Z_H = Z - Z_O = Z_O\left(\frac{1 + \dfrac{D \cdot (Z' - Z'_o)}{Z_O \cdot f}}{1 - \dfrac{Z_O \cdot (Z' - Z'_o)}{D \cdot f}} - 1 \right) \tag{8}$$

$$\text{calibration factor } Z = \frac{\text{reference body height (mm)}}{\text{measured pixel height}} \tag{9}$$

The calibration factor results from the number of pixels, which are collected by the sensor and the height which is entered by the user at the beginning of the program. The calibration factor for Z:
  The one-dimensional view above of Z-coordinates is also valid for two-and three-dimensional views as well as for X and Y coordinates.

2.4 Filtering the Pixels

While performing the measurement it may happen that slight disturbances as well as very high outliers arise, which lead to incorrect values. The outliers can come up because of a malfunction in the detection algorithm of the laser line (which is searching the columns for the pixel with the highest brightness). If some pixels are substantially brighter than the reference body, they could be detected as an edge of the reference body. To reduce interference, an appropriate environment and a measurement filter are necessary for the calibration. An appropriate environment is a place which is shielded from ambient light. Likewise all other elements (apart from the reference body) should be dark and dull, so that they reflect as less light as possible (ideally no light). A good measurement filter for outliers is the median filter. This filter sorts a series of a measurement by value and uses the value in the middle of a certain number of values as the average. Another measurement filter is the mean filter. This filter calculates the average of a particular measurement series as the arithmetic mean of the series hence the result is influenced by outliers. The median filter does not regard the discarded values any longer and they are not used to calculate the average output. I.e. a median filter with 17 elements, filters up to eight widths of pixel interferences.
  Another filter is the morphology filter. The morphology filter detects and reduces the interference and before it can influence the laser line. This filter considers each pixel individually and compares it with its neighboring pixels. The assessment can be executed as shrink (compared with 1 0, follows from 0) and as spread (compared with

1 0, it follows 1). If a median filter and morphology filter both are available, the morphology filter should be preferred to the median filter.

Another option to enhance accuracy is the aperture setting of the camera, but this can influence the measurement. The main idea is that the sensor can cover outlier values if the image is blurred. The disadvantage of this solution is that the more blurred the image, the bigger the scattering of the laser. The result is that the camera records longer than the reference body is below it in reality. The measure filters are particularly necessary to detect the borders of the reference body, because this is where the most disturbances occur. For example, the obstruction of the laser line by the reference body.

If the ROI is limited to the size of the reference body (reducing pixel width size), interference is filtered as well.

3   The implementation

3.1   Test Setup

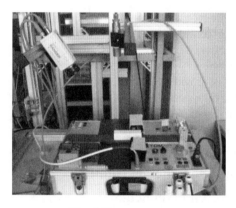

Fig. 4. Test setup

The test setup on the left side (Fig. 4.) is set up according to the "Reversed Ordinary" configuration. Before the start of the test, all devices have to be attached, the computer with keyboard and monitor (which is not visible on this picture) as well. Likewise all the caps of the sensitive components should be removed. To start the camera it has to be connected to a computer (PC) via Ethernet. There should be a compatible fast network adapter for the connection between computer and camera, so real-time communication between these two components is guaranteed. To perform the experiment the application program is started. The program communicates with the camera and analyzes the received images.

The actual evaluation of the images depends on the self-written program which provides flexible ways of measuring (e.g. for complex calibration bodies or the choice between 2D or 3D measurement). The response of the camera to the program is always the same, whereas the processing of received files is different.

3.2   Calculation of the Height and Width

The measuring of the width is realized by determining the top and bottom edges with the sensor. The width can be calculated as the difference of the two edges:

$$width = top\ edge - bottom\ edge \tag{10}$$

The reference body width, which is queried at the beginning of the program, will be divided by the measured pixel width. This division result the calibration factor for the X-coordinates.

$$\text{calibration factor } X = \frac{\text{reference body width (mm)}}{\text{measured pixel width}} \qquad (11)$$

The height is measured with a similar principle. With the measured width the columns, where the reference body is located (averaging provision) can be determined.

At this position the shades of gray with and without reference body are compared to each other. The difference between the two values is a measure for the height of the reference body. The safest choice for a column is to select the column in the middle of the reference body.

$$\text{arithmetic mean} = \frac{\text{top edge} + \text{bottom edge}}{2} \qquad (12)$$

$$\text{height} = \text{grey level with object} - \text{grey level without object} \qquad (13)$$

This formula works, if the gray level with reference body is brighter than without the reference body. Note that the shade of gray is provided in a resolution of subpixels. Because of that, the result must be divided by 16 to obtain the row value. The height of the reference body, as well as its width, is queried at the beginning of the program. This is, just as the width, divided by measured pixels. The result is the calibration factor for the Z-coordinates. The core of the 2 D calibration method was taken from [6].

$$\text{calibration factor } Z = \frac{\text{reference body height (mm)}}{\text{measured pixel height}} \qquad (14)$$

The length of the reference body can be determined by the frame counter. Therefore, the calculated values of the width and height calibration are used. The frame counter counts the number of measurement data, where in the middle position the gray level is equal to reference body. At the beginning of the program the frame counter is set to zero and for each matching profile it is incremented. The numbers of frames depend on the speed of the conveyor belt and the length of the reference body. The functionality can be tested with different lengths. The calibration factor for the Y coordinates can be calculated likewise the calibration factors for width and height.

$$\text{calibriation factor } Y = \frac{\text{reference body length (mm)}}{\text{measure number of frame}} \qquad (15)$$

At the end of the program the determined calibration factors are delivered and the calibration is now completed.

3.3 Condition of the Calibration Body

The reference body needs to meet certain qualities or conditions, so that a calibration is possible. The reference body must have a reflective or shiny surface, so that enough light is reflected to enable the sensor to recognize the reference body. A suitable color for the reference body is a bright color (for example white) which supports the reflection of light. The calibration procedure which is presented in this paper needs a rectangular reference body for 3D calibrations. 2D calibrations (Z-and X-coordinates) are possible with other reference bodies, for example triangles. A sphere is an inappropriate reference body, since it is difficult to move on the conveyor belt and it is impossible to define edges. Furthermore, note that the width of the reference body must be inferior to the width of the conveyor belt, so that the Y-plane can be measured. The more complex the geometry of the reference body is, the more flexible and comprehensive the program must be.

4 Example for the 3D Calibration

4.1 Experiments for a Reference Body

The following example calculations illustrate the above described calibration process. This is done with the "Reversed Ordinary" configuration and at an angle α=40°. The rectangular reference body used in this case, has a width of 20 mm, a height of 20 mm and a length of 180 mm. The dimensions of the reference body are entered by the user to the program. The program determines lower and upper edges; in this case, the bottom is in column 542 and the top is in column 843, hence the width is 301 columns.

The result of the gray level measure with reference body is 5871 rows. The value is divided by 16, because of the subpixel resolution. So the real value is 367 rows. The height is determined in the measurement mode and results the height in a gray level. For the illustration is here the gray level as row to understand.

The gray level without reference body is 3375, which has to be divided by 16 as well. Therefore, the gray level without reference body is 211 rows. The measured height arises to 156 rows.

The frame counter provides 331 frames for the length, thus resulting in the following calibration factors for X, Y and Z coordinates.

$$width \ = X \ = \frac{20 \ mm}{301 \ column} = 0,0664 \ \frac{mm}{column} \tag{16}$$

$$length \ = Y \ = \frac{180 \ mm}{331 \ frame} = 0,5438 \ \frac{mm}{frame} \tag{17}$$

$$height \quad = Z \ = \frac{20 \ mm}{156 \ row} = 0,128 \ \frac{mm}{row} \tag{18}$$

The following section discusses the impacts of various configurations between camera and laser on the test results. At first the basic configuration is set up. Therefore

- Reversed Ordinary(X = 1) with α=40°,
- Specular (X = 2) with α=40° and β=40° and
- Ordinary (X = 3) with β=40°, are looked at.

The best results for height, width and length provides the "Reversed Ordinary" configuration. That is shown in figure 5.

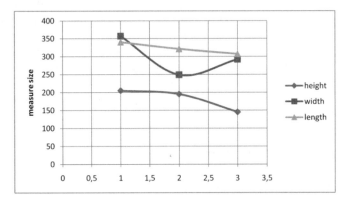

Fig. 5. Diagram of the difference between "Reversed Ordinary", "Ordinary" and "Specular"

The "Specular" configuration is suitable for height and length measurements, but provides no accurate results for width measurement. The "Ordinary" configuration turns out to be accurate for width measurement, but it contains inaccuracies/discrepancies in the height measurement.

This realization leads to the question what angle α enables the best result with the "Reversed Ordinary" configuration. Therefore the height is measured out of different angles. As a result, note that the optimum is reached between 40°<α<50°.

$$\Delta Z \approx \frac{\Delta X}{\sin(\alpha)} \qquad (19)$$

This can be mathematically proved if the series of measures for the different angles is available or the amount value for the function dependent on the angle α is derived.

The maximum can be detected between the two intersections of sine and cosine. This means exactly at 45°. From the resulting measurement the optimal arrangement of laser and camera is determined. The conclusion is that "Reversed Ordinary" with α=45° provides the best results. This measurement refers to the height and width of the reference body. ΔZ is already the estimated derivation of the triangulation law (see above). To find the maximum of ΔZ, the function needs to be derived again.

$$\Delta Z' \approx \frac{-\Delta X \cos(\alpha)}{\sin^2(\alpha)} \qquad (20)$$

By measuring the length with the frame counter another conclusion can be drawn:

The length depends on the distance between camera and reference body. The used camera was supplied with a lens, which requires a minimum distance of 0.5 m to the reference body for a sharp image. The depth of field is mainly responsible for a focused or out-of-focus image. More Information on this can be found in [8].

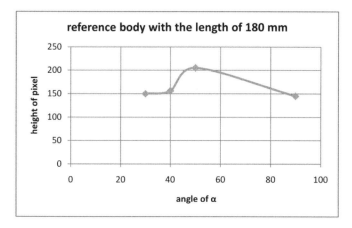

Fig. 6. Diagram of the height with difference angle

Otherwise the reference body is not sharp and there occurs a spread/dispersion which stems from the laser. If that happens the sensor detects the reference body sooner and longer.

	down	middle	top
focus	211	189	184
out of focus	170	183,5	198

Fig. 7. The table displays the differences in measurement results employing a focused or unfocused lens and various distances between camera and reference body (top=long distance (40cm); middle=middle distance (30cm) and down=small distance (20 cm))

That is why the frame counter counts more frames at the same distance and angle than whereas using a sharp lens. This result is proven by the following table and graph.

The experiment with the "Ordinary" configuration is done with short, medium and long distance. The graphic shows that the degree of sharpness plays a role depending on the distance (top = 1; mid = 2; Bottom = 3) between camera and reference body.

This phenomenon is related to the scattering of laser light and the intensity of the collected light. If the camera is close to the reference body, more frames are captured with a sharp lens than with an unfocused lens.

The scattering can become large, when the reflected light is not bright enough. Thus, the sensor detects the reference body too late. If the lens is sharper at a small distance, there is a higher interactivity of light and it can be captured earlier from the camera. In this case the image of the white reference body has a lighter shade of gray which is a problem as well.

The radiation of the unlit reference body decreases, the further the camera is moved away from the reference body. In the middle position of the camera, it is irrelevant whether the lens is set for a sharp image or not. The sensor can capture the

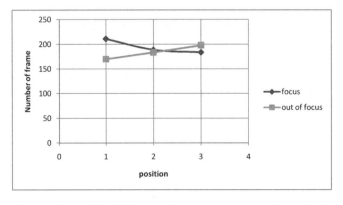

Fig. 8. The diagram shows the differences in measurement result employing focused or unfocused lens and various distances between camera and reference body (top=long distance (40cm); middle=middle distance (30cm) and down=small distance (20 cm))

laser light well and cannot be dispersed by a strong influence. The dispersion is clearly perceived when the camera is further away from the reference body, since its own radiation through the white, shiny surface of the reference body has (almost) no impact anymore. That means the laser line is detected much faster, despite the wide scatter, because the environment is much darker.

5   Conclusion

The described measurements to determine the calibration factors depend on many different factors. Only some of these factors can be influenced by the user. The user is responsible for the environment and the chosen reference body. If the environment is too bright, it can happen that the upper and lower edge swap with each other. It can also happen that the reference body gets a dark gray value instead of a bright one. If this problem occurs, upper and lower edge should be swapped with each other. Another influence on the width measurement is the position of the reference body on the conveyor belt. Depending on the angle of the reference body to the camera the sensor measures an incorrect width. Because of the lights in the environment and the position of the reference body the measurements/calibrations have slight differences. The slight differences stay in most cases in the defined tolerance range. For the measuring of the y-coordinates it is important that the shade of gray is changing depending on the position of the camera. Therefore the gray level has to be adapted to the height of the reference body, if this is not already integrated in the program. The frame number (the Y-coordinate counter) is additionally dependent on the set of triggers and the exposure time. The minimal exposure time and cycle time is 3000 μs [7]. There can be erroneous measurements, although there are special filters in place. E.g. a disturbance has been detected at the upper and lower edge. This can occur if the disorder has more than 8 pixels. If there is an error in the measurement, it must be identified and eliminated. Sometimes it helps to try another calibration or a restart.

The best reference body to choose is a pyramid. As in the geometry of a pyramid all three coordinates change the calibration is rather difficult and can thus be used for a number of bodies. For the calibration described here, a 3 D measurement is only working when the height and width do not change. The frame counter is not working without the exact pixel position of height and width.

To sum up the calibration and measurement experiments, laser triangulation for all coordinates can be carried out with the basic triangulation equations. The program should gradually be developed respectively up-graded for complex calibrations, thus enhancing flexibility.

Before the start of the calibration it should be considered, which coordinates can be determined most easily and how the results are relevant for the following problems/coordinates.

References

1. SICK IVP, Technische Beschreibung Sick IVP RANGER (2008)
2. SICK IVP, Ranger E/D Reference Manual (2008)
3. SICK IVP, FAQ Summary - 3D Cameras (2008)
4. SICK IVP, Ranger and Ruler Training Hands-On Workbook (2007)
5. SICK IVP, Ranger E/D Operating Instructions (2006)
6. Peternek, P.: 3D Laser Messsystem Ranger E55, Seminar paper, Univ. of Cooperative Education Mannheim (2008)
7. Seither, M.: Inbetriebnahme eines 3D Laser Messsystems. Seminar paper, Univ. of Cooperative Education Mannheim (2007)
8. Jennrich, O.: Schärfentiefe-,Abbildungsmaßstab- und Nahlinsenrechner (2008) (unpublished)
9. SICK IVP, Approximations of Height Resolution given Specific (2007)
10. SICK IVP, ICON Wrapper Library (2008)
11. SICK IVP, Presentations of Ranger and Ruler Technical Training (2007)
12. SICK IVP, Subpixel Distortion (2001)
13. Wikipedia contributors, Depth of focus, Wikipedia, The Free Encyclopedia (2008),
    http://en.wikipedia.org/wiki/Depth_of_focus
14. Müller, D.: Programmierung eines Laser Messsystems. Seminar paper, Univ. of Cooperative Education Mannheim (2007)

Camera-Based Control for Industrial Robots Using OpenCV Libraries

Patrick A. Seidel and Kay Böhnke

Baden-Württemberg Cooperative State University (DHBW) Mannheim, Coblitzweg 1-7,
68163 Mannheim
patrick.a.seidel@gmail.com
University of Applied Science Heidelberg
kay.boehnke@gmail.com

Abstract. This paper describes a control system for industrial robots whose reactions base on the analysis of images provided by a camera mounted on top of the robot. We show that such control system can be designed and implemented with an open source image processing library and cheap hardware. Using one specific robot as an example, we demonstrate the structure of a possible control algorithm running on a PC and its interaction with the robot.

Keywords: Robot, Control, Camera, OpenCV.

1 Introduction

In the last few years, visual object recognition and object localization have become a huge field for research and development in many different areas, especially in robotic automation [1]. The connection between robots and visual object localization has not come up unanticipated: Factory automation is largely based on robots which can almost (inter-)act without any external input or manual handling. One of the ultimate goals of industrial automation is to replace the human workforce in places where human intelligence and creativity are not needed, but accuracy, sensibility or reliability, which can be provided by a machine. In all these aspects, robots are ahead of human workers. But there is still a huge problem today's robotized automation faces: It cannot react as fast and clever as a human when it comes to the matter of flexibility. The slightest change of situation may cause a robot to be unable to proceed with its intended action. Nowadays, robots are not designed to flexibly identify and react to new situations. In fact there are some solutions provided by the manufacturers of industrial robots. They include camera based control, object recognition or object quality inspection. But these solutions are mostly expensive and specifically tailored to one type of robot. Therefore, the aim was to create a control and communication platform for industrial robots based on open source libraries, which can be easily modified, extended or upgraded.

Furthermore, it should include the basic features of industrial robot control which consist in the ability to be accurate in movement and other action, the ability to react fast and the ability to maintain a certain state of readiness, waiting for any instruction.

A. Gottscheber, D. Obdržálek, and C. Schmidt (Eds.): EUROBOT 2009, CCIS 82, pp. 144–157, 2010.

The proposed control system is able to control an industrial robot through software, which can analyze and interpret patterns made by a camera. There are various requirements imposed on this kind of control, due to its possible use in automation:

— Precise control of the robot – which is extremely important in situations involving robots working with dangerous or sensitive goods

— Fast communication between the control system (meaning the implemented software) and the robot (the hardware)

— Flexible handling of non-usual occurrences, realized through software

Admittedly, the main problem still is the exact positioning of the robot's extremity according to the evaluation of the camera pattern. The professional industrial solutions use e.g. conveyor tracking or product tracking to enable the robot / industrial machine to perform any action on the product as accurate as planned.

This paper describes the main idea, realization and further extensions of camera-based control for industrial robots using an open source library for image processing (Intel's OpenCV). In this case the realization is borne to the industrial robot Yaskawa SV3 with Motoman XRC control unit, but can be easily applied to almost any other industrial robot with a personal computer interface such as EIA-232, USB or FireWire.

The outline of the present paper is the following:

To begin with, the overview depicts the main ideas and the basic principles in conformity to which these ideas can be realized. The next part describes the algorithms used for image processing. They consist of various functions taken from Intel's OpenCV library. Further on the realization is presented using the example of the Motoman XRC industrial robot. This section includes the software and interface implementation. Following the description of the realization, a conclusion treating the results and the relevance of the project is drawn, including possible extensions or improvements.

2 Overview

2.1 Main Idea

The principal idea how to control an industrial robot via a camera is the following:
Please see Fig 1 for the visualized control scheme.

A camera needs to be mounted in a central position on the robot. This ensures that the video stream or picture taken by the camera reflects as nearly as possible the robot's view. Given that, the stream/picture (in the following referred to as picture) can be used to analyze the robot's environment and react to that environment.

The camera image is then acquired by software with an interface enabling the software to gain access to the camera. Once the image data is received, the software starts interpreting it. This step includes the analysis of the received picture. The most important parts of this analysis are the object recognition, identification and localization. The first step -Object recognition- means in this case, that the software recognizes any object on the picture that causes change.

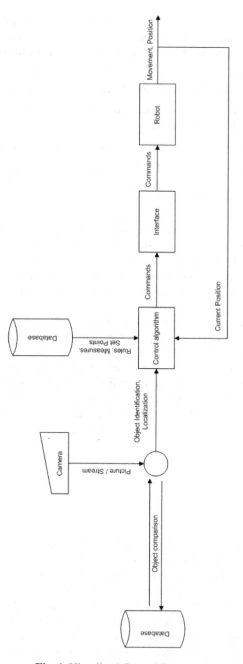

Fig. 1. Visualized Control Scheme

Assuming there is an object, an appropriate reaction can only be taken when the type and characteristics of the object as well as its location are established. Therefore object identification and object localization are needed. The identification is based on comparison of the recognized object with an object database where all known object are stored with their specific properties.

If there is a match object specific rules or actions, which are defined in the control algorithm are completed. According to the parameters defined above (such as object, position etc.) *set points* for the principal robot movement parameters are set. These are e.g. speed of movement, absolute or axis position. Before the algorithm is performed, the current position of the robot should be known to make sure, that the control algorithm does not suggest any actions that the robot is unable to execute due to singularities or axis limits.

If there is no match between the recognized object and the object stored in the database, special cases are needed to react correctly. The result of the control algorithm consists of commands for the robot, including new positions, how to get there and what to do afterwards. These commands are transmitted to the robot via common PC interfaces as mentioned above.

The actual movement of the robot has to be monitored which complies with the position monitoring. Fast and smooth movement is possible with constant variance comparison. Thus, it is ensured, that the robot reaches its destination as quickly and smoothly as possible.

2.2 Object Recognition, Identification and Localization

The main principle of object recognition is to register that the picture has changed. Once this event has occurred, object identification can be started. There are two approaches as how to identify objects:

— Extraction of object specific features followed by a comparison of these features with object features from the database

— Immediate comparison of the whole object (which has to be defined first) with some templates stored in the database

The second method is less likely to find matches, though, without defining a threshold to decide if the match is still valid.

Depending on the type of the identified object various object localization techniques are used:

The position of elliptic objects, including circles or spheres, is explicitly defined by the position of its center, therefore searching the center is the adequate measure.

The definition of a reference point is the most commonly used technique for polygonal objects. This reference point (e.g. the midpoint of one side) enables the software to allocate one position to the object in reference to which the other points of the object are described.

2.3 Control Algorithm

The control algorithm represents all the functions implemented in the robot control. According to this algorithm, with its rules, the robot reacts to "new" situations.

Parameters, which are passed to the algorithm, are the type, the position of the object and the robot's current position. The type of the object is important to determine the treatment of the object (e.g. how to pick it, where to place it). To ensure low latency periods, the control algorithm has to be processed very quickly.

The simplest realization is the calculation of the position error during tracing an object. Note that in this example, real world coordinates (mapped in a 3D-Cartesian coordinate system) are used:

$$\Delta x = x_{ObjectPos.} - x_{CurrentPos.} \tag{1}$$

$$\Delta y = y_{ObjectPos.} - y_{CurrentPos.} \tag{2}$$

$$\Delta z = z_{ObjectPos.} - z_{CurrentPos.} \tag{3}$$

2.4 Interface Communication

The robot has to communicate with the control software via standard communication interfaces. Therefore, attention has to be awarded to the correct choice of the communication interface, which hugely influences the latency periods. Delays have to be avoided wherever possible.

The communication consists of two different kinds of messages:

1. Command messages

    (Movement) Commands are sent by the control algorithm to the robot

2. Status messages

    Messages with status reports (basically position data) are sent by the robot to the control algorithm, which considers them in the course of the calculations.

As position capture is generally realized with absolute encoders, the positions registered by the encoders need to be transformed into data that can be interpreted by the robot and vice versa.

2.5 Movement

The resulting movement performed by the robot according to the transmitted commands from the control algorithm is executed as circular motion of the robot's different axes. This implies that, as mentioned in D., absolute positions have to be converted into axis positions. This calculation of robot forward kinematics parameters [12] takes some time which creates a certain latency period between the reception of the movement command and the execution. Because of this the minimum time span between the iterative command transmissions should be chosen carefully, so the latency period can be by-passed to execute the previously processed movement.

2.6 Database Integration

Databases can be integrated in different ways:

The easiest realization is to completely include the data sets in the application code. This leads to a lot of code lines but avoids an interface to an external database, which makes this realization easy to implement and fast to execute. The disadvantage is that the slightest modifications in the object properties or rules cause code changes.

The alternative is to store the data in an external database which can be accessed by the application via interface (e.g. Open Database Connectivity ODBC). This allows property or rule modifications without code changes. The implementation of the standardized interface takes some time but is not difficult to use. The additional interface is time consuming during runtime, but current database implementations process very efficiently.

2.7 General Approach

With these prerequisites the general approach for the realization is the following:

— Programming a software application that is able to control the robot in a PC environment via common PC interfaces (e.g. EIA-232, USB or FireWire)

— Embedding the image processing functions to enable object recognition, identification and localization (using Intel's OpenCV library).

— Embedding the databases with the essential object properties and rules.

— Linking the main function, the image processing and the databases to get one application that controls the robot as described beforehand.

3 OpenCV Image Processing

Intel's OpenCV (Open Source Computer Vision) library is written in C/C++ and provides basic and skilled functions for industrial image processing. For the purpose of object recognition, identification and localization, this section presents relevant functions used in the project. For more and detailed information, please refer to [2].

3.1 Preliminary Considerations

The basic considerations are taken from [3]. In previous tests, the algorithms with the best results in object recognition and localization have been found. Therefore, the functions and algorithms recommended by [3] are used in this project. A general overview of other techniques for object recognition and tracking shows [4].

3.2 Possible Methods

One of the most acknowledged functions for object tracking is the CamShift function depicted in [5]. Since this is a very flexible function which finds the object center with no concern of the object's form, a possible implementation of this function might lead to an improvement.

Another interesting approach is the use of the generalized Hough-transformation. The Hough-transformation is able to find pre-specified object forms and their centers in any image. The use and capabilities of this well-known technique are described in [6].

3.3 Applied Function

As explained in [3] the best result for object recognition and localization is achieved by comparing the camera image to template images stored in a database. The OpenCV function *cvMatchTemplate* realizes this functionality. It implements a set of methods for finding the image regions that are similar to the given template. Similarity of two images can be calculated according to correlation principles such as squared pixel difference or normalized squared pixel difference, using common mathematic approximation methods. Please refer to [2] for more information on the various approaches.

The principle used in this project is the normalized squared pixel difference method according to:

$$S(x,y) = \frac{\sum_{y'=0}^{h-1}\sum_{x'=0}^{w-1}[T(x',y')-I(x+x',y+y')]^2}{\sqrt{\sum_{y'=0}^{h-1}\sum_{x'=0}^{w-1}T(x',y')^2\sum_{y'=0}^{h-1}\sum_{x'=0}^{w-1}I(x+x',y+y')^2}} \tag{4}$$

In (4) S is the similarity between the template and the image (height h and width w) at the specific point determined by the pixel-pair (x,y). T is the pixel value of the template in the location (x,y) whereas I is the pixel value of the image in the location (x,y). The values of the similarity function are displayed in a so-called result image at their location dependent on (x,y). Fig. 2 shows the result image of cvMatchTemplate, performed with a red circle as template. This result can be used to determine the center of the recognized object, using either the above mentioned Hough-transformation or the maximum similarity value which occurs comprehensibly most of the time at the object's center location because this is the locality with the highest number of surrounding points with comparatively high similarity values. This

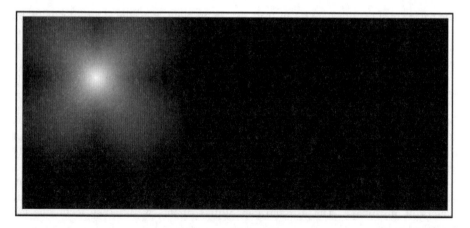

Fig. 2. Result image cvMatchTemplate

localization is only two-dimensional. A third dimension can be added, if necessary, to the localization algorithm by defining in the object properties a standard value for the size of an object. By comparing the size of the object in the database to the size of the recognized object, it is possible to calculate the distance respectively the variance from the standard distance using basic mathematics (cf. [7]).

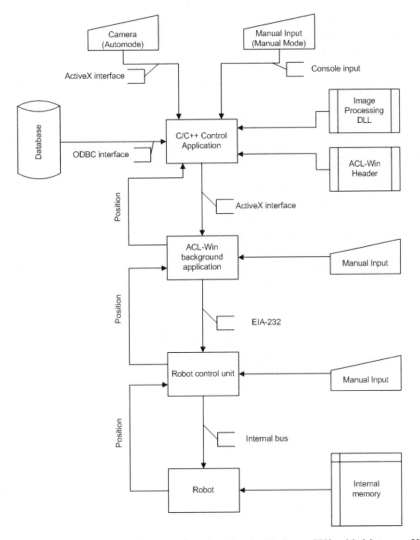

Fig. 3. Flow diagram of the robot control system for the Yaskawa SV3 with Motoman XRC control unit

4   Realization

The realization of the formerly described ideas was performed using one specific industrial robot, the Yaskawa SV3 / Motoman XRC as an example. Although the claim for universality still is very important, proving the serviceability of the approach is one first step, before conversion to other robot types respectively platforms can be faced.

The Yaskawa SV3 is an industrial robot with seven axes which can be controlled independently from each other. Programs which consist of predefined movements or actions can be stored in an internal memory. Additionally the robot can be controlled manually via the Motoman XRC robot control unit and an attached assist device. With the assist device and the control unit, the user can choose between the internally stored programs. Possibility to connect the control unit to a PC is given by using the EIA-232 interface. This permits communication between PC and robot via the control unit and therefore the control of the robot by the PC. The interface between the user and the robot is an application called ACL-Win. ACL-Win establishes the connection and provides the communication services. During all interactions between PC and robot, ACL-Win needs to be running.

For the purpose of programming C/C++ - applications with ability to communicate with the Motoman XRC, the manufacturer provides C/C++ - libraries which are easily included in any programming environment. These libraries make the programming process easy because there are multiple interfaces to overcome and to combine with just one application as shown in Fig.3.

In fact, the control application needs to concatenate very different aspects and interfaces. The most important function is the communication with the robot. As said before, ACL-Win-headers provide the basic functionality to obtain useful results and are therefore used. In the scenario depicted in Fig.3, ACL-Win and the control unit merely serve as a communication channel. The commands are sent by the control application. Almost as important as the robot communication are the acquisition and interpretation of the camera image. The control application has to be able to communicate with an attached camera and to call the image processing functions which include OpenCV functionality. When the control algorithm is implemented, it can be defined as a sort of automatic mode, where the robot reacts automatically to external events. To avoid software errors in the automatic mode, a separate manual control mode has to be available. In automatic mode, access to a database is desired, to get information for object identification and rule sets.

The next sections describe step by step the measures taken to realize the above presented robot control system.

4.1 Enabling PC Robot Control

Basic information on the PC control with the Motoman XRC can be found in [8]. Other helpful literature concerning this subject is [9] and [10]. The main feature of the PC-driven control system is that it enables the monitoring and control of the robot employing user-specific applications. In this case a console application.

4.2 Embedding Image Processing Functionality

The image processing functions which base on OpenCV methods are embedded in a dynamic link library (DLL). A DLL has multiple advantages, the most important of

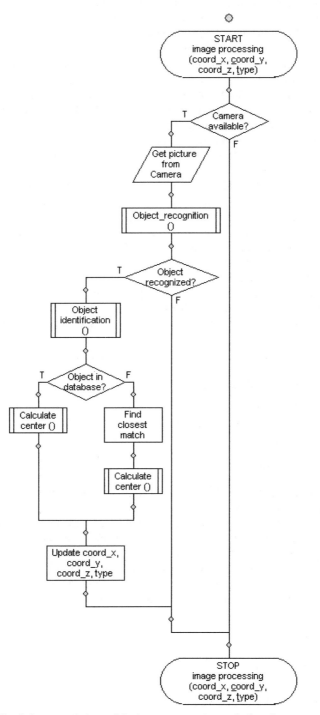

Fig. 4. Structured chart of the image processing main function

which is that it can be changed respectively maintained independently from the main application's code. Note that the image processing functionality can also be implemented using a static library or header files. In all cases, it has to be taken care of the one definition rule (ODR) which states, that any function, type or object cannot have more than one definition. The main functionality realized in the image processing DLL is the object recognition, identification and localization.

The structured chart in Fig. 4 shows the chain of events in this function.

The object identification is realized with a threshold value which has to be reached to declare an object as identified. The relating function returns a similarity value in percent, which is then compared to the threshold value. The threshold value for the identification should at least be 50 percent. The very same principle is used for the closest match algorithm.

Once the center of an identified object has been calculated, the difference from the current position of the robot's tool has to be determined. The equations (1), (2), and (3) realize this request. With the difference known, it is possible for the robot to execute the desired movement to the center of the identified object. Certainly, the robot could also process the absolute position data, but the processing of minor differences, which the deltas (increments or decrements of the position coordinate) represent, takes less time.

To obtain a true image of the robot's perspective, the camera is mounted behind the robot's gripper. This ensures that the image reflects as accurately as possible what is in front of the gripper, which is necessary to focus on objects which need to be picked by the robot. The camera is connected to the PC through USB which guarantees fast data transmission and universal connectivity.

4.3 Embedding the Database

By the deadline, no complete data set has been integrated in a database. The tests have been run with data placed in the application code. Later on a possible implementation is outlined.

4.4 Final Assembly

To get a working system, the individual parts have to be put together. Additionally, some other features have to be added. To join the different aspects means integrating them into one application. The most simple and less computing time consuming approach is a console application, which was therefore chosen. Once all parts function together, features like the position capture can be implemented. As the position capture rests upon the monitoring of robot property values, it can be realized contrarily to the transmission of commands from the PC to the robot via the control unit, so that already programmed code can be reused.

5 Conclusion

5.1 Results

To portray the results arisen whereas testing the system, the most important but also most critical points are reviewed:

− Response time

The observed response time is extremely slow (currently at ~2sec), which is by far to slow for satisfy the requirements of industrial automation. This disadvantage is caused by the communication model of the Motoman XRC. The used EIA-232 interface slows down the communication significantly. An essential gain of performance can be achieved with the use of faster or less interfaces. Faster interfaces include e.g. EIA-485, USB or Ethernet for the PC-control unit communication or field buses such as ProfiBus for the communication between control unit and robot. Another bottleneck road is the pseudo-simultaneous processing of database access, image reception and processing and communication, realized with threads. This can be converted to real-simultaneous processing using multi-core processors or computer-clusters thus accelerating response time. Evidently, the frame rate of the utilized camera influences the response time as well. The faster the camera the better gets the response time.

− Object identification accuracy

The object identification and the closest match strategy show good results even in environments with difficult or changing lighting conditions, which may occur in the case of the robot arm casting a shadow on the object whereas moving. To obtain suiting results, it is necessary to have precise templates which guarantee correct comparison. Given the fact that two very similar objects are stored (e.g. circle and ellipse) and occur with resembling properties (color, size) the employed algorithm tends to misinterpretation and takes sometimes the wrong decision. An improvement of the object identification is possible by the use of more skilled algorithms as described in III instead of the admittedly fairly simple template match function.

− Object localization accuracy

In the field tests, object localization of circular objects was successful most of the time. No tests with other object forms have been held yet. Note that it is rather facile to calculate the center of circular objects and that no explicit conclusion concerning rectangular or polygonal objects can be drawn. Possible improvement focuses on more suitable but complex algorithms using the Hough-transformation or eigenspace representation.

5.2 Relevance

It has been shown, that the proposed system realized as described above is able to control an industrial robot with cheap and available hard- and software. Since there are still some problems to deal with (analyzed in 5.1), the system is not ready to use for industrial automation applications yet, but the main idea may be enhanced and adopted to other robot types by open source software developer and therefore become ready to use. A possible outcome of this work in the near future might be a competitive visual control system for industrial robots which could especially be interesting for small and medium-sized businesses.

5.3 Extensions

During the work on this project, diverse ideas have come up concerning improvements and upgrading. The most probable ones are listed below.

— Integrating database storage

The object template pictures are stored in a certain directory and their data path is stored in a database with ODBC interface. Supplementary information such as the name of the object and important properties (e.g. size, color etc.) are also stored in the database. Implementation of the template database is in the development stadium using a MySQL database and the corresponding MySQL Connector /ODBC driver [11]. The idea to store control algorithm rules likewise in a database has been abandoned as these rules can be realized in functions in either the control application itself or in the robot's internal memory and therefore be easily maintained.

— Remote access

Nowadays almost every PC is able to communicate via internet. No wonder that some industrial applications, notably monitoring, are running online on a server connected to the internet or another network and are hence accessible from everywhere in the world (or at least from everywhere in the network). The same concept could be adapted to the control system: a remote monitoring feature of the robot. Additionally the manual mode as well could be integrated network-based.

— Wireless camera communication

Almost unlimited mobility is a feature that enables modern robots to meet even requirements such as extreme flexibility. Mounting a wired camera on the robot's gripper could definitely limit the robot's mobility. To avoid that, a possible solution might be the use of wireless protocols to establish the connection between PC and camera. Examples are Bluetooth, ZigBee or Wi-Fi, which all derive from the IEEE standards 802.11 and 802.15.

— Teaching mode

The integration of a teaching mode which offers the possibility to directly "teach" the robot new objects could be a huge step into the direction of autodidactic learning. In this mode, a beforehand unknown object is put in the field of view in a standard distance and by a manual command, the object's image and its properties are stored in the template database. Afterwards, predefined rules can be applied to the new object.

Acknowledgment

P. A. Seidel would like to thank K. Böhnke for his contribution to this paper, T. Schubert and B. Thomas for their excellent prior work on that subject, U. Schulz for sharing the laboratory understandingly and G. Christoph, G. Becker, and F. Madjzoub for their help and support.

References

1. Corke, P.I.: Visual Control of Robot manipulators – A Review. In: Hashimoto, K. (ed.) Visual Servoing: Real Time Control of Robot Manipulators Based on Visual Sensory Feedback, pp. 1–31. World Scientific, Singapore (1993)
2. Intel Corp.: Open Source Computer Vision Library – Reference Manual (2001)

3. Schubert, T.: Industrielle Bildverarbeitung mit Hilfe einer Open Source Bibliothek in C++ (unpublished)
4. Lordemann, C.G., Lambers, M.: Objekterkennung in Bilddaten (unpublished)
5. Bradski, G.R.: Computer Vision Face Tracking For Use in a Perceptual User Interface. Intel Technology Journal, Intel Corp. (1998)
6. Duda, R.O., Hart, P.E.: Use of the Hough Transformation to Detect Lines and Curves in Pictures. ACM Comm. 15(1), 11–15 (1972)
7. Arendt, P., et al.: Roboter-Kommunikation mit externen Systemen, p. 30 (unpublished)
8. Motoman Corp.: Motoman XRC – Reference Manual (2000)
9. Kirchgessner, R.: Inbetriebnahme einer Steuerungsschnittstelle des Yaskawa SV3 unter C++ (unpublished)
10. Wegmann, L.-O.: Steuerung eines Industrieroboters unter C++ (unpublished)
11. DuBois, P.: MySQL, ch. 6. Addison-Wesley, Reading (2008)
12. Denavit, J., Hartenberg, R.S.: A Kinematic Notation for Lower-Pair Mechanisms Based on Matrices. ASME Journal of Applied Mechanics, 215–221 (1955)

Generating Complex Movements of Humanoid Robots by Using Primitives

Miomir Vukobratović[1], Branislav Borovac[2], Mirko Raković[2], and Milutin Nikolić[2]

[1] Institute Mihajlo Pupin,Volgina 15, 11000-Belgrade, Serbia
vuk@robot.imp.bg.ac.yu
[2] University of Novi Sad, Faculty of Technical Sciences, 21000-Novi Sad,
Trg Dositeja Obradovića 6, Serbia
borovac@uns.ns.ac.yu, rakovicm@uns.ns.ac.yu,
milutinn@ uns.ns.ac.yu

Abstract. The objective of this work is to demonstrate the possibility of using primitives to generate complex movements that ensure motion of bipedal humanoid robots. Primitives represent simple movements that are either reflex or learned. Each primitive has its parameters and constraints that are determined on the basis of the movements capable of performing by a human. The set of all primitives represents the base from which primitives are selected and combined for the purpose of performing the corresponding complex movement. The proof that a correct selection of primitives is made and that the movement is the appropriate one is obtained on the basis of the maintainance of dynamic balance, which is realized by monitoring the ZMP position, as well as based on the pattern of the very movement.

Keywords: humanoid, pattern generation, dynamic balance, ZMP, primitives.

1 Introduction

In the realization of their movements, many humanoid robots use predefined reference motions [1-4], the main goal in their realization being to prevent fall, i.e. to preserve dynamic balance, and then, realize the intended movement in a most faithful way. In this work we will focus our attention on the realization of bipedal gait. First, based on the predefined motion that satisfies all preset kinematic and dynamic requirements (this motion is called *reference* motion), control quantities are generated that will be forwarded to the actuators to realize the reference motion. By including control we ensure the elimination of errors arising due to the ever-present disturbances. It is generally known that bipedal robots are very sensitive to disturbances, and that maintenance of dynamic balance represents the primary control task. In [4,5], different control algorithms were analyzed based on classical PID regulators for preserving an anthropomorphic motion, whereas a separate control law was defined, based on the deviation of the ZMP from its reference position, for preserving dynamic balance. The notion of dynamic balance has been considered in detail in [6,7], especially from the aspect of biological principles that are used to

A. Gottscheber, D. Obdržálek, and C. Schmidt (Eds.): EUROBOT 2009, CCIS 82, pp. 158–172, 2010.

preserve it, as well as in the case of a nonstandard foot-ground contact. In [8], a comparison was made of the fuzzy logic control and PID control. The mentioned strategies were mainly adequate for compensating small disturbances, whereas the compensation of large disturbances requires a different approach. The problem of classifying disturbances as small and large ones has been considered in detail in [9].

In the compensation of large disturbances (e.g. when the robot is pushed by a large force or when it kicks its foot against an obstacle) the dynamic balance is instantaneously jeopardized; hence the system has to fully abandon the realization of the reference motion and undertake an urgent action (impose new control signals), to preserve dynamic balance, i.e. to prevent the fall. For example, an urgent action may be a step aside if the disturbance was a strong lateral force. Only after reestablishing dynamic balance, the system can return to the reference motion and its further realization by imposing the reference control.

A main problem in compensating for large disturbances in the described way is the extremely short time available to form a compensating movement, since the shortage of time does not allow realization of some complex computer and time demanding algorithms. Hence, in this work we propose an approach that enables on-line generation of compensating movements for bipedal robots, based on using simple primitives, which are relatively easier to modify. In [10-14], the authors used the entire movement as a primitive (overall gait, transition from standing to walking, etc.). An essential difference between such approach and the one proposed in this work is that the on-line motion is formed by a combination of primitives and not of the movements recorded in advance. Complex movements were decomposed into simple movements, called primitives (e.g. leg stretching, leg bending, hip turning, etc.). The idea is based on the fact that a human, even when dynamic balance is not directly endangered, realizes reflexly a series of simple movements and changes his motion in real time. Thus, for example, if in the course of the gait a need arises to modify the motion (e.g. there appears an obstacle that is to be bypassed or the need to walk the staircases) the human does not waste time on calculation of a new and complex motion to be realized and of new kinematic and dynamic parameters of motion, but it selects the most appropriate primitive from the already lerned ones, by adjusting only some of the basic parameters such as the height of the leg lifting, angle of leg bending or the stride length. By introducing the base of primitives that are realized by taking as a model human's movements, the aim of this work was to demonstrate that the appropriate selection and combination of primitives can yield the realization of a complex motion that was not planned in advance, and, at the same time, preservation of dynamic balance.

In this work we will focus on the basic explanations of the notion and forms of primitives, and the approach will be illustrated on the example of gait realization in the absence of disturbances. Further development of this approach, especially for the case of occurrence of large disturbances, will be reported in our forthcoming articles.

The article describes first the model of the humanoid robot used for the simulation. This is followed by a description of the primitives to be realized and the way they were used in the gait realization. Finally, we present simulation results, obtained on the basis of the presented approach to generating complex movements using primitives.

2 Model of the Humanoid Robot and Notion of ZMP

We will present first the kinematic structure of the humanoid robot used in the present work. The software used had the capability of forming a dynamic model of the humanoid robot composed of a number of open and closed kinematic chains. Use was made of the software that is based on the concept of a free-flying mechanism [15], which consists of one or more kinematic chains whose links are interconnected by revolute joints with one degree of freedom (DOF). One of the mechanism links is the basic link, which branches into kinematic chains (Fig. 1).

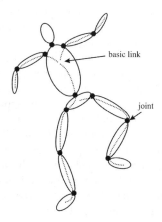

Fig. 1. Free-flying mechanism

The state in which the free-flying mechanism is observed is described by the vector Q, which consists of the position (x, y, z) and orientation of the basic link (φ, θ, ψ) and the vector q of n elements, determining the relative angle between two adjacent links, where n represents the number of the mechanism links. So, the total number of the mechanism DOFs is 6+n.

$$Q = \left[x,\ y,\ z, \varphi,\ \theta,\ \psi,\ q_1,\ q_2, \ldots, q_n \right]^T \tag{1}$$

In order to realize the mechanism motion it is necessary that a driving torque acts at each active joint, realized by the corresponding actuator, so that we have n driving torques. Driving torques can not be applied to the first 6 DOFs that determine the position and orientation of the basic link, so that the vector of driving torques is defined in the following way:

$$T = \left[0,\ 0,\ 0, 0,\ 0,\ 0,\ \tau_1,\ \tau_2, \ldots,\ \tau_n \right]^T \tag{2}$$

In accordance with the defined vectors Q and T, the dynamic model of the humanoid robot is given the equation:

$$H(Q) * \ddot{Q} + h\left(Q, \dot{Q} \right) = T \tag{3}$$

The mechanism used in this work possesses four kinematic chains and 50 DOFs (see Fig. 2). The first and second kinematic chains form the left and right leg of the humanoid, the third represents the trunk and the right arm, while the fourth one stands for the left arm.

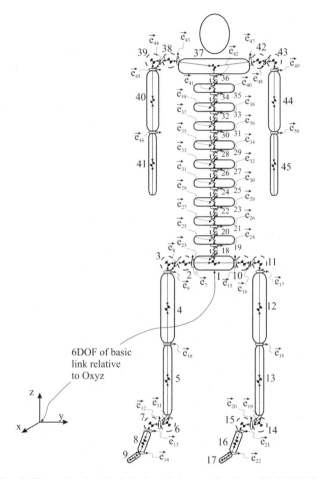

Fig. 2. Kinematic sketch of the free-flying mechanism with 50 DOFs

The left and right legs consist of the pelvis, upper leg, lower leg, foot body and the toes link (the foot consists of two links), so that the kinematic chain of the legs consists of 16 DOFs. The left and the right arm consist of the clavicle, upper arm, and the lower arm. The kinematic chains of the arms have in all 8 DOFs. The trunk consists of 10 links interconnected by two joints whose axes are mutually orthogonal, and theyensure the rotation about the x and y axes. Thus, the trunk is segmented and allows motion like that of a human with the backbone. The trunk has a total of 20 DOFs.

Particular links correspond to the real mechanism links (link 5 to the shank, link 4 to the thigh, etc.), and are presented in Fig. 2 by full lines. However, some links were needed only for the purpose of modeling the joints with more DOFs. Namely, the joints with more DOFs are modeled as sets of joints having only one DOF each, and which are connected by links having mass, moment of inertia and length equal to zero (in Fig. 2 being presented by dashed lines). Thus, for example, the hip joints, which are in reality spherical joints with three DOFs, are modeled as sets of three one-DOF joints whose axes are mutually orthogonal. Thus the right hip is modeled by a set of simple joints 7, 8 and 9 (with the unit vectors of rotation axes e_7, e_8 and e_9), and the left hip by the set of joints 15, 16 and 17 (unit vectors e_{15}, e_{16} and e_{17}). The links connecting these joints (for the right hip the links 2 and 3, and for the left links 10 and 11 were needed only to satisfy the mathematical formalism of kinematic chain, on which the software is based) were presented by dashed lines, to indicate their "fictitious" nature. The other links (those that are not part of the joints with more DOFs) correspond to the real characteristics of the links of an average human body.

The trunk is not realized as a single link but as a 10-link one. Each of the joints connecting the trunk links has 2 DOFs and each of them enables relative rotation about the axes that are oriented in the directions of the x and y axes.

The software used for modeling and simulation of the humanoid robot motion has the possibility of realizing and disrupting contact between the free-flying mechanism and surrounding objects. Thus, for the purpose of gait simulation an object is introduced that represents the ground on which the robot moves. In our case this was a flat surface. During the walk the robot is always in contact with the ground, at least via one foot. When the robot is in the single-support phase the contact exists between the supporting leg and the ground, and in the double-support phase both feet are in contact with the ground. At the moments of transition from the single-support to the double-support phase and vice versa, realization and disruption of the contact takes place, respectively.

Evidently, the humanoid model considered here is very complex, especially from the aspect of the multilink trunk. We decided to use such a complex model because we expect it will be very convenient and indispensable in the subsequent phases of research, although the additional complexity of the trunk has no essential influence on the results presented in this paper.

The basic task of locomotion systems is locomotion itself. While walking locomotion systems are freely supported by their feet on the ground but do not overturn. In such case we say they are dynamically balanced. An indicator of dynamic balance is Zero Moment Point (ZMP). To define the ZMP consider a Cartesian coordinate frame with the x and y axes being tangential to the flat ground and the z axis being normal. In case humanoid perform walk (do not overturn) during either single or double support phase of walking there is always an unique point inside support area for which $\Sigma M_x = 0$ and $\Sigma M_y = 0$, where M is moment about an axis generated by the ground reaction forces. This point is called ZMP.

3 Experimental Results

In this chapter we will present the primitives and describe the way of their use. Then we will demonstrate the possibility of using the primitives introduced to realize complex movements. Simulation of bipedal motion using the primitives will be described and the simulation results will be presented.

3.1 Primitives

As already mentioned, the term primitive stands for a simple reflex or lerned movement that a human or robot is capable to realize. A primitive itself should be simple in order it could be easily combined with the other primitives. Each primitive is parametrized and has the following parameters:

- intensity of the movement in the span of 0-1, which determines the extent to which, for example, a leg is to be bent or streched,
- time instant during which the primitive is executed, and
- duration of the primitive, i.e. the time in which the primitive is to be executed.

Fig. 3 a) shows stick diagrams representing bending (at the hip, knee and the ankle) of the leg that is in swing phase. Therefore, by imposing the requirement

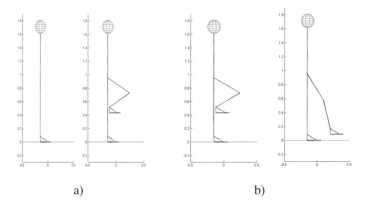

a)   b)

Fig. 3. Stick diagrams of the humanoid robot model in the realization of primitives by the swing leg: a) leg bending, b) leg stretching

(by imposing the primitive) that the robot is to bend the leg that is in swing phase, the leg is lifted the way presented in Fig. 3 a). Fig. 3 b). shows the primitive by which the bent leg in swing phase is stretching, whereby use is made of the joints at the hip, knee, ankle and the link of the toes of the leg being in swing phase. This primitive is realized so to produce an appropriate stretching of the leg which will become supporting leg in the beginning of the double-support phase.

Each of the primitives is realized by activating one or more DOFs. Thus, for example, the primitive for bending the leg in swing phase involves activation of the joints at the hip, knee and ankle of the swing leg. Fig. 4 shows the primitives at each of these joints for the case of leg bending presented in Fig. 3 a). In this case the magnitude of the angle at the knee is twice larger than, and is of an opposite sign to, the magnitudes of the angles at the hip and the ankle. It should be noticed that in this

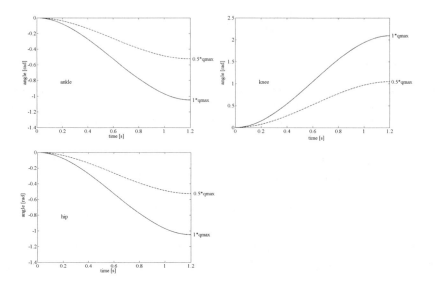

Fig. 4. Change of the angles at the ankle, knee, and the hip during swing phase in the case of leg bending

case the angle of the toes link remained constant all the time and equal zero; hence its diagram has not been shown. Also, it is worth noting that the primitives can be changed very easily in the sense of varying the range of the changing angle (by multiplying with a factor smaller than 1 the span of the angle change will be narrower compared to q_{max}, where q_{max} represents the maximal value of the angle allowed by the structure of the given joint, so that there is no sense to multiply the q_{max} by a factor that is greater than 1), as well as by changing the duration of its realization (faster or slower movement execution).

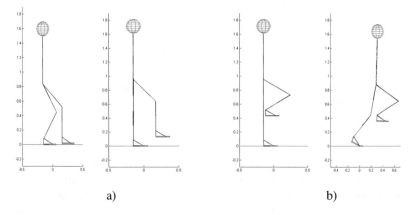

a)   b)

Fig. 5. Stick diagrams of the model of humanoid robot in the realization of primitives by the supporting leg: a) stretching, b) inclining, by which the system as a whole moves forward

Apart from the primitives imposed onto the swing leg, the primitives acting on the supporting leg: stretching of the supporting leg (Fig. 5a), as well as the primitive for the mechanism's inclination forward (Fig. 5b), were also realized.

In the realization of the primitive for stretching of the supporting leg (Fig. 5a) use is made of the ankle, knee, and hip joints. By the realization of primitives for stretching the supporting leg these joints move so to ensure that at the end of the movement, when the stretching intensity was preset to 1, the leg is fully stretched, which corresponds to the angles at the joints of 0 rad. Of course, it is possible to change the intensity of leg stretching, by which is changed the span of the motion at the joints, and, by the same token, the characteristics of the movement.

In the realization of the primitive for inclining, the supporting leg is activated again and the same joints are used as in the realization of the primitive for stretching the supporting leg (Fig. 5b), but now the toes link is also activated since the rear part of the foot (heel) separates from the ground and the system remains supported only on the link of the toes. With this primitive, the angles at the joints of the knee and hip are of the opposite sign to the angles at the links of the toes and ankle, to ensure keeping the trunk in the upright position with respect to the external coordinate frame.

3.2   Simulation Results

In assessing the quality of motion realized by applying primitives the gait synthesized by the semi-inverse method [1-4] served as a reference. In this method, the motion of the legs[1] and the ZMP position are given in advance, and then the trunk motion is synthesized so to ensure realization of the prescribed ZMP position. Since the ZMP is located in the predicted position inside of the support area, the robot is dynamically balanced during the whole half-step. Fig. 6 shows the stick diagrams of the reference motion during a half-step, while Fig. 7 pictures the reference trajectory of the ZMP.

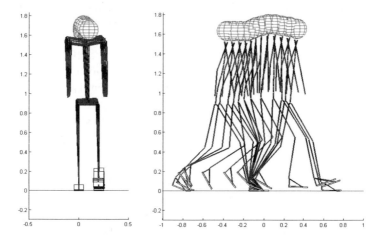

Fig. 6. Stick diagrams of dynamically balanced reference half-step synthesized by semi-inverse method

---

[1] The motion of the legs was obtained based on human walk.

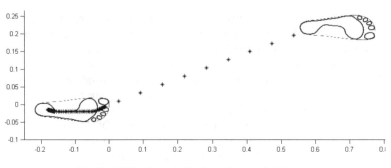

Fig. 7. ZMP trajectory for the reference half-step

The example on which we will illustrate the realization of primitives is the gait that aims at being anthropomorphic. In order to realize it, the reference motion at the joints by which the robot moves forward (the joints corresponding to the rotation about the y axis, i.e. the joints 9, 10, and 14 for the right leg, and the joints 17, 18 and 21 for the left leg) is replaced with the corresponding primitives. The motion of the legs joints whose orts are parallel to the x and z axes and all motions of the trunk and arms were taken over from the reference motion. The state identical to the initial state in the reference motion was taken as the starting state. All movements presented lasted 2.4 s, which was realized in the simulation at the sampling interval of $\Delta t = 0.00066$ s, corresponding to a total of 3600 iterations.

The newly obtained motion of the legs consists of the following combination of primitives. As first, bending of the swing leg was imposed an intensity of 0.5 (dashed line on the diagrams in Fig. 4). The starting moment of the swing leg bending was the beginning of the movement (first iteration), and the primitive duration was 1800 iterations. At the same time, the primitive for stretching (straightening) of the supporting leg was imposed on the intensity of 1, which lasted 1200 iterations. This movement was followed by the primitives of the swing leg stretching and inclination of the supporting leg. Stretching of swing leg involved the following parameters: intensity 1, the primitive was imposed on the 1800th iteration on, and lasted the next 1800 iterations, i.e. to the end of the movement. The intensity of inclination of the supporting leg was 1, the primitive was imposed starting from the 1200th iteration and lasts to the end of the movement, i.e. the next 2400 iterations. In this way we obtained a movement that lasted 3600 iterations (like the reference one), and the result of the applied primitives was the movement represented by the stick diagrams in Fig. 8.

Fig. 9 shows the magnitudes of the angles at the ankle, hip and the knee of the supporting leg for the reference movement and for the movement obtained by using primitives for one half-step. It should be noticed that the motion in all presented diagrams for the supporting leg was a result of the combination of two primitives taking place in the sequence, i.e. stretching and inclining.

The imposing of the new, changed, motion, formed on the basis of primitives at the prespecified joints while keeping the reference motion at the other joints, yielded a

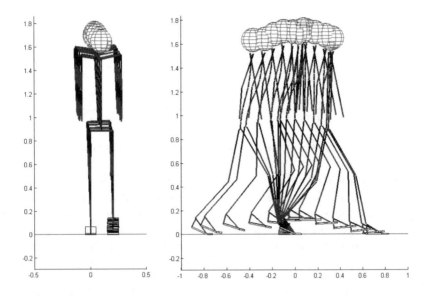

Fig. 8. Stick diagrams of the half-step when the motion was realized with the aid of primitives

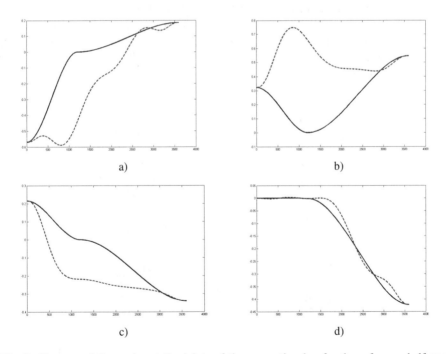

Fig. 9. Changes of the angles at the joints of the supporting leg for the reference half-step (dashed line) and the half-step obtained on the basis of primitives (full line): a) hip, b) knee, c) ankle, d) toes

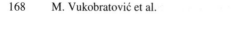
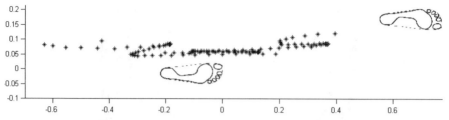

Fig. 10. ZMP trajectory for the half-step obtained by imposing primitives onto the legs

robot, the fall would be inevitable. Hence it would be necessary to correct the motion imposed at the joints, in order to preserve dynamic balance. The correction was realized in the following way. Since the general pattern of motion was satisfactory it was to be changed to a smallest possible extent, and it was only necessary to change the ZMP position. Because of that it was decided to change the acceleration at the ankle of the supporting leg in each sampling period to a sufficient extent so that the ZMP would be brought sufficiently close to its reference position (an acceptable deviation from the reference position was prescribed, and, when the ZMP entered that zone, the task was considered fulfilled). Double integration of the corrected accelerations yielded the corrected trajectories at the ankle.

Fig. 11 shows the stick diagrams of a half-step of the humanoid robot when the motion of the legs was realized by using primitives and applying correction at the ankle of the supporting leg, while Fig. 12 illustrates the ZMP trajectory in the case of the corrected motion.

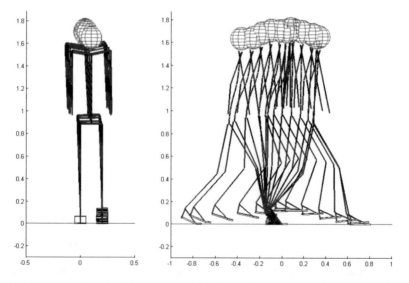

Fig. 11. Stick diagrams of the half-step when the primitives are imposed onto the legs, with the correction at the ankle

Magnitudes of the angles about the x and y axes with and without correction at the ankle of the supporting leg are presented in Fig. 13 c) and d). It should be noted that significant distortion of the ZMP trajectory, as shown in Fig. 10. As can be seen, the robot is not dynamically balanced, and if such motion would be imposed onto a real

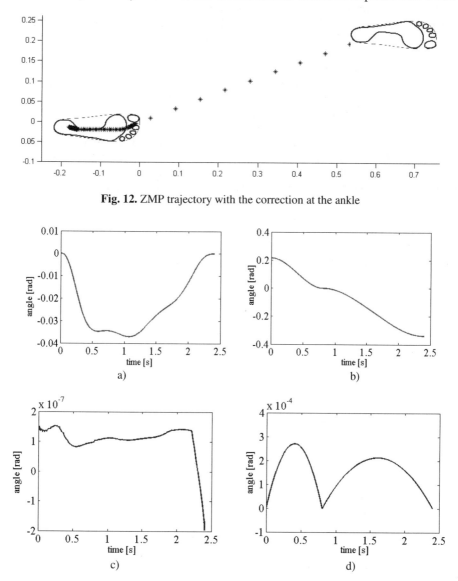

Fig. 12. ZMP trajectory with the correction at the ankle

Fig. 13. Magnitudes of the angles with and without correction at the ankle at which correction was performed and the difference arising due the correction: a) angle about the x axis at the ankle with and without correction, b) angle about the y axis at the ankle with and without correction, c) the difference of the angles about the x axis of the ankle arising due to the correction, d) the difference of the angles about the y axis arising due to the correction applied at the ankle.

both curves in Fig. 13 a) and b) coincide with each other, so that they are practically seen as one curve. As can be seen from Fig. 13 c) and d), the deviations at the ankle before and after the correction are minimal and did not introduce significant change, as far as the movement pattern is concerned. Maximal difference between the angles at the ankle before and after the correction does not exceed the value of $3*10^{-4}$ rad. By comparing Fig. 11 and Fig. 13 we can also see that the motion did not change significantly and the desired pattern of the half-step has been preserved. However, a comparison of Fig. 10 and Fig. 12 reveals that the application of the correction yielded an essential improvement in the ZMP trajectory, and it is very alike the reference trajectory of the ZMP shown in Fig. 7. This correction at the ankle, which, as already mentioned, did not influence the movement pattern, yielded a successful realization of a dynamically balanced gait using primitives.

A question arises as to whether the correction of primitives the way demonstrated in this work would be applicable if the robot had to perform on-line modification of its motion. We believe that this would be possible, for at least two reasons. As first, in the on-line modification of the motion it is not necessary to change completely the primitives imposed onto the legs, but only modify some of the parameters such as, for example, extending the stride, lifting the leg a little bit higher to surmount the obstacle that appeared in the way, or rotating additionally the leg at the hip to bypass the obstacle. The other reason is the fact that the correction can be applied in the form of position control of the ZMP (correction of small disturbances), by which dynamic balance would be preserved, as was demonstrated in our previous work [5].

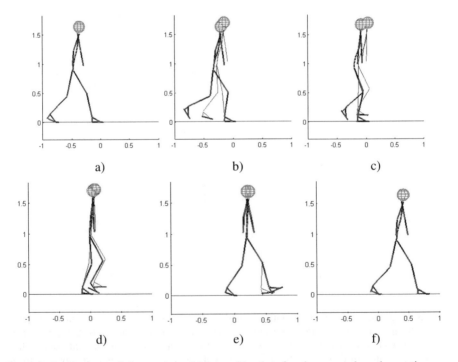

Fig. 14. Comparison of the motion of humanoid robot for the case when the motion was synthesized by the semi-inverse method (full line) and with the aid of primitives (thin line)

Let us discuss briefly the comparison of the characteristics of the gaits synthesized by the semi-inverse method (Fig. 14, full line) and by applying primitives (Fig. 14, thin line). The figure shows the superimposed stick diagrams for the two synthesized motions. It is clear that the postures at the beginning and at the end of motion coincide, which was a consequence of the appropriately prescribed limiting conditions, but between the limiting points the system did not move in a completely identical way in the two cases. This we do not consider as a shortcoming but as a potential advantage, because there is no any objective reason for the system to move in the same way. However, it should be pointed out that the system during the entire motion was dynamically balanced, and that there was no threat of falling down at any time instant. We expect that the results would be also similar in the case of an on-line modification of the motion.

4 Conclusion and the Directions for Further Work

Humanoid robots of the future will have to move in an unstructured environment, so that it will not be possible to plan and generate the entire robot's motion in advance. The direction and motion parameters will have to be determined and modified during the motion realization. Hence the need to develop a new approach that will enable modification of the current motion, or the synthesis of a completely new one in real time, will be an inevitability.

In this work we presented one of the ways in which solving of this problem can be approached. In contrast to the approaches in which the motion of the overall locomotion system is generated as a whole, we propose to compose the entire motion of the locomotion system of a series of motions at the particular joints. If a need appears that the humanoid has to adapt to the situation it found itself in the given moment, it is possible to simple modify the motion at the particular joints (e.g. to make a turn in order to bypass an obstacle), since the motion at each joint represents a separate primitive. It is of special interest to notice that the combination of several simple primitives can lead to a significantly more complex movement.

The objective of this work was to clearly explain the basics of the notion, form and application of primitives, so that this work is based on the realization of the gait of humanoid in the absence of disturbances. The basis and the reference with respect to which we considered the deviations of the motion realized with the aid of primitives served the gait synthesized by the well-known semiinverse method. In order to demonstrate the use of the new approach the motion at paticular joints that was synthesized by the semiinverse method was replaced by primitives, whereas the motion of the rest of the system remained unchanged. Because of this intervention, the system's dynamic balance was completely disturbed, but, by applying a small correction at the ankle of the supporting leg, we succeeded to correct the motion, so that dynamic balance was fully ensured during the whole gait sequence.

We expect that the approach presented here could be applicable for both on-line modifications of motion and generating compensating motions in the case of large disturbances, when the system is in danger of falling down. In the future work we will intensify our efforts in dealing with this problem, as well as in the application of primitives on the motions of the other joints of the locomotion mechanism.

References

1. Vukobratović, M.: How to Control the Artificial Anthropomorphic Systems. IEEE Trans. on System, Man and Cybernetics SMC-3, 497–507 (1973)
2. Vukobratović, M.: Legged Locomotion Systems and Anthropomorphic Mechanisms Mihajlo Pupin Institute, Belgrade (1975), also published in Japanese, Nikkan Shimbun Ltd. Tokyo, MIR, Moscow (1976) (in Russian), Beijing (1983) (in Chinese)
3. Vukobratović, M., Borovac, B., Surla, D., Stokić, D.: Biped Locomotion – Dynamics, Stability, Control and Application. Springer, Berlin (1990)
4. Vukobratović, M., Borovac, B.: Zero-Moment Point- Thirty Five Years of its Life. Int. Jour. of Humanoid Robotics 1(1), 157–173 (2004)
5. Vukobratović, M., Borovac, B., Raković, M., Potkonjak, V., Milinović, M.: On some aspects of humanoid robots gait synthesis and control at small disturbances. Int. Jour. of Humanoid Robotics 5(1), 119–156 (2008)
6. Vukobratović, M., Herr, H., Borovac, B., Raković, M., Popovic, M., Hofmann, A., Potkonjak, V.: Biological Principles of Control Selection for a Humanoid Robot's Dynamic Balance Preservation. Int. Jour. of Humanoid Robotics 5(4), 639–678 (2008)
7. Vukobratović, M., Borovac, B., Potkonjak, V., Jovanović, M.: Dynamic Balance of Humanoid Systems in Regular and Irregular Gaits: an Expanded Interpretation. Intl. Journal of Humanoid Robotics 6(1) (in press, 2009)
8. Vukobratović, M., Borovac, B., Raković, M.: Comparison of PID and Fuzzy Logic Controllers in Humanoid Robot Control of Small Disturbances. In: Eurobot Conference 2008, pp. 69–79. Matfyz Press, Heidelberg (2008)
9. Vukobratović, M., Borovac, B., Potkonjak, V.: Towards a Unified Understanding of Basic Notions and Terms in Humanoid Robotics. Robotica 25, 87–101 (2007)
10. Hauser, K., Bretl, T., Latombe, J.-C.: Using Motion Primitives in Probabilistic Sample-Based Planning for Humanoid Robots. In: Algorithmic Foundation of Robotics VII, vol. 47, pp. 507–522. Springer, Heidelberg (2008)
11. Zhang, L., Bi, S., Liu, D.: Dynamic Leg Motion Generation of Humanoid Robot Based on Human Motion Capture. In: Proceedings of the First International Conference on Intelligent Robotics and Applications Part I, pp. 83–92. Springer, Heidelberg (2008)
12. Denk, J., Schmidt, G.: Synthesis of walking primitive databases for biped robots in 3D-environments. In: Proceedings of the IEEE International Conference on Robotics and Automation (ICRA), Taipei, Taiwan, vol. 1, pp. 1343–1349 (2003)
13. Nakaoka, S., Nakazawa, A., Yokoi, K., Ikeuchi, K.: Leg Motion Primitives for a Humanoid Robot to Imitate Human Dances. Journal of Three Dimensional Images 18(1), 73–78 (2004)
14. Ha, S., Han, Y., Hahn, H.: Adaptive Gait Pattern Generation of Biped Robot based on Human's Gait Pattern Analysis. International Journal of Mechanical Systems Science and Engineering 1 1(2), 80–85 (2007)
15. Potkonjak, V., Vukobratović, M., Babković, K., Borovac, B.: General Model of Dynamics of Human and Humanoid Motion: Feasibility, Potentials and Verification. Int. Jour. of Humanoid Robotics 3(2), 21–48 (2006)

Author Index